Lecture Notes in Statistics

Edited by S. Fienberg, J. Gani, K. Krickeberg, I. Olkin, and N. Wermuth

Shelemyahu Zacks

Stochastic Visibility in Random Fields

›ringer-Verlag
·w York Berlin Heidelberg London Paris
‹yo Hong Kong Barcelona Budapest

Shelemyahu Zacks
Department of Mathematical Sciences
Binghamton University
PO Box 6000
Binghamton
New York 13902-6000

Library of Congress Cataloging-in-Publication Data Available
Printed on acid-free paper.

Camera ready copy provided by the author.
Printed and bound by Braun-Brumfield, Ann Arbor, MI.
Printed in the United States of America.

9 8 7 6 5 4 3 2 1

ISBN 0-387-94412-5 Springer-Verlag New York Berlin Heidelberg

Dedicated to The Memory of Professor Micha Yadin

Preface

The present monograph is a comprehensive summary of the research on visibility in random fields, which I have conducted with the late Professor Micha Yadin for over ten years. This research, which resulted in several published papers and technical reports (see bibliography), was motivated by some military problems, which were brought to our attention by Mr. Pete Shugart of the US Army TRADOC Systems Analysis Activity, presently called US Army TRADOC Analysis Command. The Director of TRASANA at the time, the late Dr. Wilbur Payne, identified the problems and encouraged the support and funding of this research by the US Army. Research contracts were first administered through the Office of Naval Research, and subsequently by the Army Research Office. We are most grateful to all involved for this support and encouragement.

In 1986 I administered a three-day workshop on problem solving in the area of sto-chastic visibility. This workshop was held at the White Sands Missile Range facility. A set of notes with some software were written for this workshop. This workshop led to the incorporation of some of the methods discussed in the present book into the Army simulation package CASTFOREM. Several people encouraged me to extend those notes and write the present monograph on the level of those notes, so that the material will be more widely available for applications.

The original planning of Professor Yadin and mine was to write this monograph jointly. Unfortunately, Professor Yadin died in April 1991, after a long illness, and we had not had the chance to write the monograph together. I dedicate this monograph to his memory.

I would like to thank again Mr. Pete Shugart, for his continued interest, encourage-ment and support. Without his enthusiasm and dedication much of this would not have been achieved. I would like to thank the Directors of the Mathematics and Statistics Sections of the Office of Naval Research and the Army Research Office for their sup-port. I would like to thank also the administration of Binghamton University for their support. Last, but not least, I would like to thank Mrs. Marge Pratt for her excellent typing of the manuscript.

Zacks
Binghamton, NY
August, 1994

Table of Contents

troduction

. Aims and Objectives

sibility problems are of much interest in military applications of operations research, communications and many other subject areas. Imagine that you are standing in a est not far from a vehicle. You see the vehicle completely. The vehicle starts to move ay from you, driving among the trees. After a short while you see only some parts the vehicle, and shortly after that you lose sight of the vehicle completely. Trees are domly dispersed between you and the vehicle and interfere with the lines of sight. is is a typical visibility problem. What is the stochastic or random elements of the blem? If you are told exactly where the trees are located, the width of their trunks l all other pertinent information, and asked what fraction of the vehicle you would able to see after it drove in a certain direction 100 meters away from you, you will be e in principle to figure this out. The problem is deterministic. On the other hand, ve do not have all the pertinent information we could figure only, assuming that the es are randomly dispersed, their sizes are random, etc., certain probabilities that the ctions of the vehicle that could be seen are of certain size. We turn the problem from a erministic problem, which requires a lot of information, that is often unavailable, to a chastic problem whose solution depends on the assumed model of randomness. The sent book provides the reader the methods of determining visibility probabilities l related distributions, assuming that the objects which obscure the visibility are domly distributed in certain regions according to a model called "Poisson random d". The Poisson fields are either standard ones, or non- homogeneous. The subject tter of the present monograph can be considered as a special (though complicated) e of coverage processes. We could model the problem as one in which a source of light ocated at the observer's location. Obscuring elements in the field cast their shadows the target. If these shadows cover the target completely, or a large fraction of it, target is not visible. The reduction of our visibility problems to coverage problems lds generally very complicated coverage processes, of random shadows which are not ependent and not identically distributed. Yadin and Zacks (1982), in their first paper the subject, treated the problem as a coverage process on the circle (see Section 2.7). er on it was realized that, for actually solving some non-trivial visibility problems it ght be more convenient to have a different approach, and to develop special tools. The sent book is aimed at presenting these tools in a manner which should be accessible to ders, who do not have the theoretical knowledge in the areas of stochastic processes l random fields. The reader who is interested in the abstract theory of coverage cesses should study the book of P. Hall (1988), which contains also a long list of erences to papers and books written on the coverage problem.

The present book contains most of the special tools and methods developed by M. lin and S. Zacks in a series of papers (see references) written, under several contracts h the Office of Naval Research and the Army Research Office (see preface). Our in objective is to present these methods in a way that will be beneficial for potential rs. For this purpose, our level of exposition is such that readers who know calculus, elementary geometry and trigonometry, should be able to study the methods and ap them. Moreover, all the algorithms are computerized. The reader can use ready ma executable programs to solve most of the problems and exercises, which are given Chapter 7. A list and short description of these programs is given in an appendix. T solutions of exercises (see Chapter 7) are given in sufficient details, to assist the rea with the material.

In the next section we provide examples of military and non-military problems t can be solved with the tools provided in the present book. In Section 0.3 we presen summary (synopsis) of the material which is discussed in Chapters 1 - 6 of the book

0.2. Some Military Applications

Many of the weapon systems require clear visibility of the target for a certain len of time. This visibility may be interrupted by objects in the field, which are part the terrain like trees, bushes, etc. The interruption might be caused by moving obje which cut the lines of sight. There are many different versions of military proble which require clear lines of sight. We present here a class of problems, which we c "the hunter-killer problem", as an example of visibility problems in a random field.

Consider a situation in which a set of M targets, e.g., tanks, move through a "cl tered" piece of terrain towards a hunter (observer), e.g., helicopter. By "cluttered" mean that, throughout the area, there are distributed (in some fashion) various obs cles that break-up the line-of-sight between the M targets and the observer. In additi there are K escorts, e.g., air defense systems accompanying the M targets. The esco typically occupy stationary positions, to the rear of the moving targets, that prov good fields of view toward likely observer positions. The single hunter occupies one of potential observation points, all of which provide "good" coverage of the area thro which the M targets must move.

This is a two-sided game, with the target playing the role of passive participat only; the escorts and the hunter being the active elements. The role of a target is to m along the assigned path at some prescribed speed. As it does, it periodically prese itself to view from one or more of the observation points, one of which is occupied the hunter (with some prescribed frequency and duration). Those intervals (of time distance) when a segment of the path can be seen from a particular observation po will be referred to as a window of visibility (between observer and target). During t interval a target moving along that path could be engaged by the hunter. The hunt however, requires some specified time (measured from the initiation of line-of-sight round impact) to complete an engagement. Therefore the size of some visibility wind may allow engagements, while others may not. Given that an engagement occurs, th is a kill probability which is a function of the range separation between target a observer.

The stationary escorts occupy covered and concealed positions that offer "go coverage of possible hunter positions. For each hunter - escort pair, the existence line-of-sight can be described. When a hunter appears at an observation point, a is detected by the escort, the latter attempts an engagement. The escort requ a certain unit of time (from start of line-of-sight to round impact) to complete engagement. Whereas, as previously stated, the observer requires some other num of time units to complete its engagement. The outcome of the escort's attempt to engage an exposed observer can be assessed with a prescribed range dependent kill probability. The questions of interest are:

a. What is the probability that an escort kills the hunter during the game.
b. Expected number of targets killed by the hunter.
c. Probability of kill by an escort given an observer presentation.
d. Probability of lost engagement opportunity due to loss of line-of-sight between observer and target.

One can generalize this class of problems to include also three dimensional surveillance problems. Aircraft or satellite fly over a region which should be surveyed by radar, infrared or other photographic equipment. Some or all of the targets on the ground might not be detected due to interfering objects, like clouds, crowns of trees, etc. The question is, what is the probability of detecting a certain number of targets?

Visibility problems exist also in naval problems. A naval target in the ocean might be obscured by an iceberg, by waves, or other possible obscuring elements.

The methods discussed in the present book could provide answers to many relevant questions concerning the visibility of targets.

There are many non-military applications of coverage theory (see Hall (1988), Solomon (1978)). The problems dealt with in the present book relate to a special type of coverage, which is induced by shadows cast by obscuring objects. Communication via satellites requires clear lines of communication between points in space and points on earth. Space is now crowded with satellites. Certain regions on earth might be obscured from a given point in space by satellites which are in the field. The methods of the present book can be applied to evaluate the probability that a line of communication will be available between two points.

0.3. Synopsis

The material of the book is presented in six chapters. Chapter 1 presents the probability models for visibility problems, a glossary of distribution functions used in this book, and a brief introduction to random fields. The exercises provide additional examples. Some readers may wish to review elementary probability theory before studying this book. We refer them to textbooks devoted to probability theory and its applications, like W. Feller, Vol. I (1968), or S. Ross (1976).

Chapter 2 introduces some classical problems of geometrical probability, and presents some elementary treatment of coverage problems. The interested reader is referred also to some classical books on the subject. In particular, see Ambartzumian (1982, 1990), Hall (1988), Kendall and Moran (1963), Santaló (1976), and Solomon (1978).

Chapter 3 starts with the topic of visibility probabilities. The model is typically that of one observer and several target points in the plane. The obscuring elements are modeled as random disks with random radii and centers which are randomly located in a region between the observer and the target points (see Figure 0.1). The number of such disks, in any given region, having certain size, is either fixed or has a binomial or Poisson distribution. Two methods are presented. The geometric method, which is appropriate for standard Poisson fields, and the analytical method, which is more general. In all cases we develop formulae for the probability of simultaneous visibility of

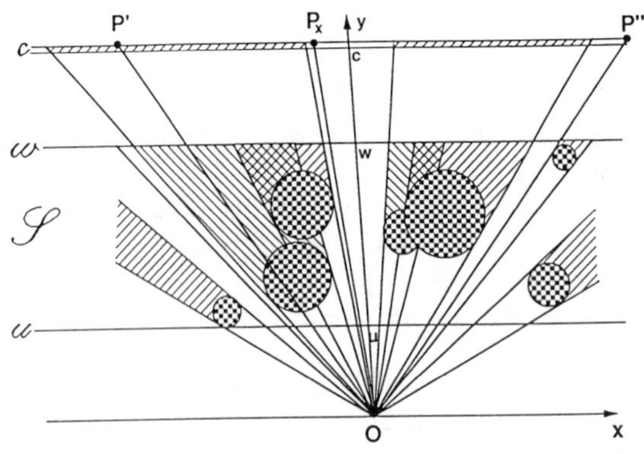

Figure 0.1. Random Field of Disks And Lines of Sight

several target points. The chapter ends with a section generalizing the methods, so that the probability of simultaneous visibility of "windows" can be determined. Windows are intervals (or neighborhoods) of certain size around the target points, which should be completely visible in order that the targets could be detected.

Chapter 4 continues with more advanced topics of simultaneous visibility probabilities of several points. We start with the problem of multi-observer multi-target visibility probabilities in the plane, and present methods, developed by Yadin and Zacks (1990), for computing these probabilities.

After dealing with visibility problems of several observers, we discuss in Chapter 4 visibility problems in three dimensional spaces, when the obscuring objects are modeled as spheres of random radius, with centers which are uniformly distributed in three dimensional layers. These methods were developed by Yadin and Zacks (1988), for solving problems connected with air defense or air attack.

Chapters 5 and 6 deal with distribution functions of certain random measures of visibility. We start in Chapter 5 with a few counting measures. For example, how many targets, out of m specified ones, can be observed simultaneously, from a given observation point. These methods are then generalized to determine the joint distributions of the number of targets observable from several observation points. The second part of Chapter 5 is devoted to the distribution of a random integrated measure, W, on star shaped curves. A special case of W is the random length of a line segment that is visible. The first step in approximating the distribution of W is to determine its moments. A recursive equation is derived for determination of the moments. From these moments we can, in principle, determine the distribution of W. A beta mixture approximation to the cumulative distribution of W is developed. We conclude the chapter by considering a discrete approximation to the distribution of the length of segments on the line which are under shadow (invisible). These methods follow the study of Yadin and Zacks (1986). Finally, Chapter 6 presents methods for determining the distribution of the length of a visible segment, to the right of a specified point, and the length of the invisible segment to the right of a point. These distributions are then applied to determine the survival probability of targets which have to cross a certain path, as a function of the length of the path. We develop there also an algorithm for the determination of the distribution of the number of invisible segments (shadows) on a line.

Chapter 7 is devoted to problems and solutions. A large number of problems illustrate the theory and its applicability.

All the material in the chapters 1-6 is illustrated with a large number of figures, and many numerical examples are presented, using the specially designed software.

1

Probability Models

In the present chapter we present the probability models which are basic to the problems discussed in the monograph. We provide also a short glossary to the distribution functions used in the following chapters. It is assumed that the reader is familiar with probability theory. The glossary is provided for establishing notation. The reader who needs additional studying of this material is referred to the books of William Feller (1968) or Sheldon Ross (1976).

1.1. Probability Models For Obscuring Elements

We consider first a stochastic visibility problem. Suppose that an observer is located at the origin $\mathbf{O}$, and looks towards a point $\mathbf{P}$. The line segment $\overline{\mathbf{OP}}$ is called a **line of sight**. The point $\mathbf{P}$ is said to be **visible** from $\mathbf{O}$ if the line of sight $\overline{\mathbf{OP}}$ is not intersected by obscuring elements. Obscuring elements are 3-dimensional bodies. In the present monograph we model the obscuring elements as **spheres**. Let (X, Y, Z, R) be the coordinates of a sphere. (X, Y, Z) are the location coordinates of its center and R is its radius. The coordinates (X, Y, Z) are generally rectangular. Sometimes it is convenient to use spherical coordinates (ρ, θ, ϕ) to present a center of a sphere. In a stochastic visibility model, the obscuring elements are **random spheres**. In this case, (X, Y, Z, R) are random variables. The stochastic model ascribes these random variables a joint distribution, from which we can deduce the probability of certain events. More specifically, if $\mathcal{B}$ denotes the Borel sigma-field generated by the random variables (X, Y, Z, R) and $F(x, y, z, r)$ denotes the joint cumulative distribution function (c.d.f.) of (X, Y, Z, R) then, the probability of a Borel set B in $\mathcal{B}$ is

$$\Pr\{B\} = \int_B dF(x, y, z, r). \tag{1.1}$$

Typically our models assume that $F(x, y, z, r)$ is absolutely continuous in all variables i.e., there exists a non-negative function $f(x, y, z, r)$ (p.d.f) such that

$$\Pr\{B\} = \iiiint\limits_B f(x, y, z, r)\,dx\,dy\,dz\,dr \tag{1.2}$$

for all $B \in \mathcal{B}$.

Let N denote the number of obscuring elements (random spheres) in the region which might intersect the line of sight. N is either fixed or random, following som distribution. For a given value $n \geq 1$, we assume that **conditionally**, given $\{N = n\}$ the random vectors (X_i, Y_i, Z_i, R_i), $i = 1, \cdots, n$, representing n random spheres, ar independent and identically distributed (i.i.d.).

Most of the material in the present monograph is on procedures to evaluate the s **multaneous visibility** from $\mathbf{O}$ of m points $\mathbf{P}_1, \cdots, \mathbf{P}_m$, $m \geq 2$. It is generally assume that $\mathbf{O}, \mathbf{P}_1, \cdots, \mathbf{P}_m$ are **coplanar**. The random spheres in space either intersect tl plane of interest or do not. The intersecting sets are **random disks**. In the followir

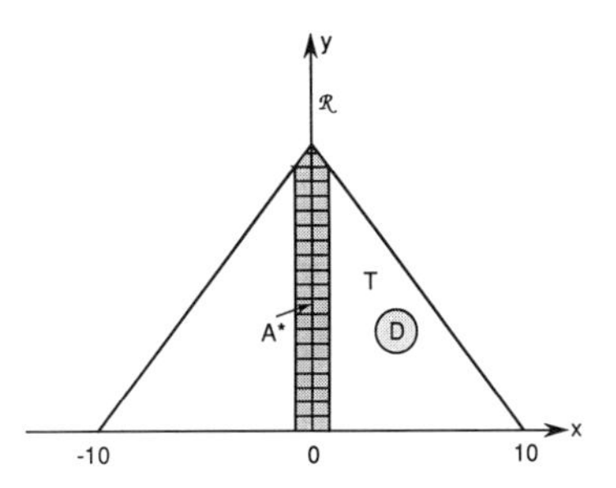

Figure 1.1. Geometric description of A within T.

example we illustrate the use of geometrical arguments to compute the probability that a random disk intersects a line of sight.

Example 1.1. Consider an equilateral triangle T in the plane whose vertices are specified by the Cartesian coordinates $(-10, 0)$, $(10, 0)$ and $(0, 10)$. [The scale is of arbitrary unit length, (see Fig. 1.1)]. The experiments consist of randomly placing the center of a disk D, of radius 0.5, inside T. What is the probability that the disk D intersects the ray $\mathbb{R}^*$ from the origin $(0,0)$ through the vertex $(0, 10)$? Since the radius R is fixed, we have to consider only two random variables (X, Y), which are the coordinates of the center of D in T.

We construct the following probability space (model). The sample space Ω is the set of all points (x, y) within T, i.e.,

$$\Omega = \{(x, y); -10 < x < 10,\ 0 < y < 10 - |x|\}.$$

Let $\mathcal{B}^2$ be the Borel sigma-field on the plane, containing all intervals of the form $(-\infty, x] \times (-\infty, y]$, $-\infty < x, y < \infty$. The sigma-field $\mathcal{F}$ we consider is the collection of all sets $B \cap \Omega$, where $B \in \mathcal{B}^2$. Since the disk center is placed at random within T, the probability function $\Pr\{\cdot\}$ assigns every (measurable) set A in $\mathcal{F}$ the value

$$\Pr\{A\} = \text{Area}\{A\}/\text{Area}\{T\}.$$

The disk D intersects the ray $\mathbb{R}^*$ if, and only if the x-coordinate of its center is within the set

$$A^* = \{(x, y) : |x| < 0.5\}.$$

Accordingly, the probability that D intersects $\mathbb{R}^*$ is

$$\Pr\{A^*\} = 0.0975.$$

A generalization of this problem is given in Example [2.3]. ■

The second example shows the construction of a distribution function (c.d.f.), which is induced from the basic model.

Example 1.2. Consider a triangular region T, specified by the vertices $(-1,0)$, $(1,0)$ and $(0, 1)$. A disk D is randomly placed in T. Let the random variables X and Y denote the coordinates of the point at which the center of the disk is placed. We develop below, by geometrical methods, the c.d.f. of the random distance, R, of (X, Y) from $(0,0)$.

Notice that if $r < 1/\sqrt{2}$ then the half disk $\{(x,y) : x^2 + y^2 \leq r^2\} \cap T$ has an area of $\pi r^2/2$. Hence $\Pr\{R \leq r\} = \dfrac{\pi r^2}{2}$ for all $0 < r < 1/\sqrt{2}$. When $\frac{1}{\sqrt{2}} < r < 1$ we have to subtract from $\pi r^2/2$ the area of $\{(x,y) : x^2 + y^2 \leq r^2, 0 < y\} - T$. Simple geometry yields that this area is equal to $r^2(\frac{\pi}{2} - 2\cos^{-1}(\frac{1}{2r}(1+\sqrt{2r^2-1}-1))) - \sqrt{2r^2-1}$. Thus,

$$F_R(r) = \Pr\{R \leq r\} = \begin{cases} \pi r^2/2, & \text{if } 0 \leq r \leq 1/\sqrt{2} \\ \\ 2r^2 \cos^{-1}(\frac{1}{2r}(1+\sqrt{2r^2-1})) + \sqrt{2r^2-1}, & \text{if } \frac{1}{\sqrt{2}} \leq r \leq 1. \end{cases}$$

Notice that $\cos^{-1}(x)$ denotes the arc cosine of x, in radians. ■

Let θ denote the probability, according to a given model, that a random disk in the plane intersects a specified line of sight. The conditional visibility probability of that point, given $\{N = n\}$, is $\psi_n = (1-\theta)^n$. The non-conditional visibility probability is $\psi = E\{(1-\theta)^N\}$, where $E\{\cdot\}$ denotes the expected value according to the distribution of N. For a model which assumes that N has a Poisson distribution (see Section 1.2), with mean μ, $\psi = e^{-\mu\theta}$. This is a basic result for Poisson random fields (see Section 2.7).

In order to compute the simultaneous visibility probability from $\mathbf{O}$ to $\mathbf{P}_1, \cdots, \mathbf{P}_m$, let $\mathcal{L}_1, \cdots, \mathcal{L}_m$ denote the m lines of sight. Let D be a random disk and let $\{D \cap (\bigcup_{i=1}^{m} \mathcal{L}_i)\}$ denote the event that D intersects at least one of the lines of sight. Let $\theta_m = \Pr\{D \cap (\bigcup_{i=1}^{m} \mathcal{L}_i)\}$. The simultaneous visibility probability is

$$\psi_m = E\{(1-\theta_m)^N\}. \tag{1.5}$$

Methods for computing ψ_m are developed in Chapter 3. Chapter 4 discusses simultaneous visibility probabilities from several observation points.

1.2. Glossary of Distributions

In the present section we list distribution functions which are mentioned in the present monograph. Generally, we denote by $E\{\cdot\}$ the **expected value** of the random variable in the brackets. $V\{\cdot\}$ denotes the **variance** of a random variable.

1.2.1. Some Discrete Distributions

1.2.1.1. Binomial Distribution

Let K_n be the number of successes among n identical and independent trials, such that the probability of success, θ, in each trial is fixed (Bernouli trials). The probabilit distribution function (p.d.f.) of K_n is

$$b(j; n, \theta) = \binom{n}{j} \theta^j (1-\theta)^{n-j}, \quad j = 0, \cdots, n, \tag{1.6}$$

The cumulative distribution c.d.f. is

$$B(x; n, \theta) = \begin{cases} 0, & \text{if } x < 0 \\ \sum_{j=0}^{[x]} b(j; n, \theta), & \text{if } 0 \le x < n \\ 1, & \text{if } x \ge n \end{cases}$$

where $[x]$ is the integer part of x. For this distribution

$$E\{K_n\} = n\theta \tag{1.8}$$

$$V\{K_n\} = n\theta(1-\theta). \tag{1.9}$$

1.2.1.2. Poisson Distributions

A random variable assuming only the non-negative integers has a Poisson distribution if its p.d.f. is

$$p(j; \lambda) = e^{-\lambda} \frac{\lambda^j}{j!}, \quad j = 0, 1, 2, \cdots. \tag{1.10}$$

The c.d.f. is

$$P(x; \lambda) = \begin{cases} 0, & \text{if } x < 0 \\ \sum_{j=0}^{[x]} p(j; \lambda), & \text{if } x \ge 0. \end{cases} \tag{1.11}$$

For a Poisson random variable

$$E\{X\} = \lambda, \tag{1.12}$$

$$V\{X\} = \lambda. \tag{1.13}$$

The following is an important result for Poisson distributions: **If $X_1, X_2, \cdots, X_k$ are independent random variables having Poisson distributions with means $\lambda_1, \cdots, \lambda_k$, respectively, then $T = \sum_{i=1}^{k} X_i$ has a Poisson distribution with mean** $\lambda_1 = \lambda_1 + \cdots + \lambda_k$.

.2.1.3. Multinomial Distributions

et $\{C_1, \cdots, C_m\}$ denote m disjoint and exhaustive sets (a partition) with respect to ome sample space. Consider n independent and identical trials and let $J_{n,i}$ denote the number of trials which result in C_i, $i = 1, \cdots, m$. Obviously $\sum_{i=1}^{m} J_{n,i} = n$. The joint distribution of $(J_1, \cdots, J_{m-1})$ is called an m-nomial distribution. Let θ_i denote the probability that a trial results in C_i, $i = 1, \cdots, m$. $\sum_{i=1}^{m} \theta_i = 1$. The joint p.d.f. of $(J_1, \cdots, J_{m-1})$ is

$$p(j_1, \cdots, j_{m-1}; n, \theta_1, \cdots, \theta_m) = \frac{n!}{\prod_{i=1}^{m} j_i!} \prod_{i=1}^{m} \theta_i^{j_i}, \quad j_i \geq 0, \sum_{i=1}^{n} j_i = n \tag{1.14}$$

where $j_m = n - \sum_{i=1}^{m-1} j_i$. For the multinomial distribution,

$$E\{J_i\} = n\theta_i, \quad i = 1, \cdots, m \tag{1.15}$$

$$V\{J_i\} = n\theta_i(1 - \theta_i), \quad i = 1, \cdots, m \tag{1.16}$$

$$\text{Cov}(J_i, J_{i'}) = -n\theta_i\theta_{i'}, \quad \text{for } i \neq i'. \tag{1.17}$$

Here Cov$(\cdot, \cdot)$ denotes the covariance of two random variables. The following is an important result for the developments in this book.

If the conditional distribution of $(J_1, \cdots, J_m)$, given N and $\theta_1, \cdots, \theta_m$, is m-nomial, and if N has a Poisson distribution with mean λ, then the non-conditional distribution of $(J_1, \cdots, J_m)$ is like that of m independent Poisson random variables with $E\{J_i\} = \lambda\theta_i$, $i = 1, \cdots, m$.

1.2.2. Some Continuous Distributions

1.2.2.1. Uniform Distributions

A distribution function is uniform (or rectangular) on an interval $[a, b]$, if its p.d.f. is constant over that interval, i.e.,

$$f(x; a, b) = \begin{cases} \dfrac{1}{b-a}, & \text{if } a \leq x \leq b \\ 0, & \text{otherwise.} \end{cases} \tag{1.18}$$

The corresponding c.d.f. is

$$F(x; a, b) = \begin{cases} 0, & \text{if } x < a \\ \dfrac{x-a}{b-a}, & \text{if } a \leq x \leq b \\ 1, & \text{if } b < x. \end{cases} \tag{1.19}$$

The expected value and variance of a random variable having such a distribution is

$$E\{X\} = (a+b)/2, \tag{1.20}$$

and

$$V\{X\} = (b-a)^2/12. \tag{1.21}$$

1.2.2.2. Beta Distributions

A random variable has a Beta distribution, with parameters ν_1 and ν_2, if it is distributed over the interval $[0, 1]$, and its p.d.f. is

$$f(x;\nu_1,\nu_2) = \begin{cases} \frac{1}{B(\nu_1,\nu_2)} x^{\nu_1-1}(1-x)^{\nu_2-1}, & 0 \le x \le 1 \\ 0, & \text{otherwise.} \end{cases} \tag{1.22}$$

The parameters ν_1 and ν_2 are any **positive** real numbers. The function $B(\nu_1, \nu_2)$, called the beta function, is given by

$$B(\nu_1,\nu_2) = \int_0^1 x^{\nu_1-1}(1-x)^{\nu_2-1}dx. \tag{1.23}$$

If ν_1 and ν_2 are positive integers, say k and l, then

$$B(k,l) = \frac{1}{(k+l-1)\binom{k+l-2}{k-1}}. \tag{1.24}$$

The c.d.f. of Beta(ν_1, ν_2) is given by

$$F(x;\nu_1,\nu_2) = \begin{cases} 0, & \text{if } x < 0 \\ I_x(\nu_1,\nu_2), & \text{if } 0 \le x \le 1 \\ 1, & \text{if } x > 1 \end{cases} \tag{1.25}$$

where $I_x(\nu_1, \nu_2)$, is called the **incomplete beta function ratio**, and is given by the formula

$$I_x(\nu_1,\nu_2) = \frac{1}{B(\nu_1,\nu_2)} \int_0^x u^{\nu_1-1}(1-u)^{\nu_2-1}du. \tag{1.26}$$

By changing variables of integration one can readily show that

$$I_x(\nu_1,\nu_2) = 1 - I_{1-x}(\nu_2,\nu_1), \quad 0 < x < 1. \tag{1.27}$$

The expected value and variance of a random variable having a Beta(ν_1, ν_2) distribution are:

$$E\{X\} = \nu_1/(\nu_1+\nu_2), \tag{1.28}$$

and

$$V\{X\} = \frac{\nu_1 \nu_2}{(\nu_1 + \nu_2)^2(\nu_1 + \nu_2 + 1)} \tag{1.29}$$

1.2.2.3. Gamma Distributions

A random variable distributed over the positive real numbers $(0, \infty)$ has a Gamma distribution with parameters β and ν, $0 < \beta < \infty$, $0 < \nu < \infty$, if its p.d.f. is

$$f(x; \beta, \nu) = \begin{cases} \dfrac{1}{\Gamma(\nu)\beta^\nu} x^{\nu-1} e^{-x/\beta}, & 0 < x < \infty \\ 0, & x < 0 \end{cases} \tag{1.30}$$

$\Gamma(\nu)$ is called the gamma function, with argument ν, and is given by,

$$\Gamma(\nu) = \int_0^\infty x^{\nu-1} e^{-x} dx, \qquad \nu > 0. \tag{1.31}$$

The following recursive formula is obtained by integration by parts:

$$\Gamma(\nu) = (\nu - 1)\Gamma(\nu - 1), \qquad \nu > 1. \tag{1.32}$$

Hence, for every positive integer k, $\Gamma(k) = (k-1)!$. The parameter β is a **scale parameter** and ν is a **shape** parameter.

The expected value and the variance of a random variable having a gamma distribution are:

$$E\{X\} = \nu\beta, \tag{1.33}$$

and

$$V\{X\} = \nu\beta^2. \tag{1.34}$$

The Gamma distribution with parameters β and ν will be designated by $G(\beta, \nu)$. The **exponential distribution** is $G(\beta, 1)$. The distribution $G(2, k/2)$, $k = 1, 2, \cdots$ is called the **chi-squared distribution** with k degrees of freedom, and is denoted by $\chi^2[k]$.

Another important property of random variables having a Gamma distribution is:

If X_1 is distributed like $G(\beta, \nu_1)$ and X_2 is distributed like $G(\beta, \nu_2)$, and if X_1 and X_2 are independent, then $T = X_1 + X_2$ is distributed like $G(\beta, \nu_1 + \nu_2)$, and $R = \dfrac{X_1}{X_1 + X_2}$ is distributed like Beta(ν_1, ν_2). **Moreover, T and R are independent.**

Accordingly, the sum of independent and identically distributed exponential random variables has a $G(\beta, n)$ distribution.

1.2.2.4. Normal Distributions

A random variable X has a Normal (Gaussian) distribution, with parameters μ, σ, if its p.d.f. is

$$f(x;\mu,\sigma) = \frac{1}{\sigma\sqrt{2\pi}} \exp\left\{-\frac{1}{2}\left(\frac{x-\mu}{\sigma}\right)^2\right\}, \qquad -\infty < z < \infty. \tag{1.35}$$

We designate this distribution by $N(\mu,\sigma)$. This p.d.f. is **symmetric** around μ, with a scale parameter σ. The expected value and variance of X are:

$$E\{X\} = \mu, \tag{1.36}$$

$$V\{X\} = \sigma^2. \tag{1.37}$$

When $\mu = 0$ and $\sigma = 1$ the distribution is called the **standard normal**. We denote by $\phi(x)$ the p.d.f. of $N(0,1)$ and by $\Phi(x)$ its c.d.f. $\Phi(x)$ is called also the **standard normal integral**.

1.2.2.5. Bivariate Normal Distributions

A pair of random variables, (X,Y), have a joint bivariate normal distribution if their joint p.d.f. is

$$f(x,y;\xi,\eta,\sigma_x,\sigma_y,\rho) =$$
$$\frac{1}{2\pi\sigma_x\sigma_y\sqrt{1-\rho^2}} \exp\left\{-\frac{1}{2(1-\rho^2)}\left[\left(\frac{x-\xi}{\sigma_x}\right)^2 - 2\rho\frac{x-\xi}{\sigma_x}\cdot\frac{y-\eta}{\sigma_y} + \left(\frac{y-\eta}{\sigma_y}\right)^2\right]\right\},$$
$$-\infty < x, y < \infty. \tag{1.38}$$

(ξ,η) is the location point of the center of the distribution, σ_x and σ_y are scale parameters, and ρ is an orientation parameter. Contours of equal p.d.f. are given by the concentric ellipses,

$$\left(\frac{x-\xi}{\sigma_x}\right)^2 - 2\rho\frac{x-\xi}{\sigma_y} + \left(\frac{y-\eta}{\sigma_y}\right)^2 = c. \tag{1.39}$$

The expected values, variance and coefficient of correlation are

$$E\{X\} = \xi, \qquad -\infty < \xi < \infty \tag{1.40}$$

$$E\{Y\} = \eta, \qquad -\infty < \eta < \infty \tag{1.41}$$

$$V\{X\} = \sigma_x^2, \qquad 0 < \sigma_x < \infty \tag{1.42}$$

$$V\{Y\} = \sigma_y^2, \qquad 0 < \sigma_y < \infty \tag{1.43}$$

and the correlation between X and Y is

$$\rho(X,Y) = \rho, \qquad -1 < \rho < 1. \tag{1.44}$$

The joint c.d.f. at a point (x, y) can be expressed as

$$F(x, y; \xi, \eta, \sigma_x, \sigma_y, \rho) = \Phi_2\left(\frac{x-\xi}{\sigma_x}, \frac{y-\eta}{\sigma_y}; \rho\right), \tag{1.45}$$

where $\Phi_2(z_1, z_2; \rho)$ is the standard bivariate normal c.d.f. at (z_1, z_2), with correlation parameter ρ. This function can be determined according to the formula

$$\Phi_2(z_1, z_2; \rho) = \int_{-\infty}^{z_1} \phi(x)\Phi\left(\frac{z_2 - \rho x}{\sqrt{1-\rho^2}}\right) dx. \tag{1.46}$$

1.3. Random Fields

In the present section we outline the stochastic structure which represents random objects scattered in space.

Let $\mathcal{S}$ be a set* in $\mathbb{R}^n$ (a line segment, a region in the plane or in a 3-dimensional space). Let $\mathcal{P} = \{\mathbf{p}_1, \cdots, \mathbf{p}_N\}$ be a finite number of points (N is fixed or random) scattered within $\mathcal{S}$. We associate with the i-th point of $\mathcal{P}$, $\mathbf{p}_i$, a random vector (variable) $\mathbf{X}_i$, which yields the location coordinates of $\mathbf{p}_i$ within $\mathcal{S}$. It is assumed that, given N, $\mathbf{X}_1, \cdots, \mathbf{X}_N$ are conditionally independent random vectors having an identical distribution, with c.d.f. $H(\mathbf{x})$, whose support is $\mathcal{S}$, i.e., $\int_{\mathcal{S}} dH(\mathbf{x}) = 1$.

We associate with each point of $\mathcal{P}$ a random variable (vector) Y, called a **marker**. This random variable may present the size or shape of the random objects centered at the points of $\mathcal{P}$. For example, in many of the applications to visibility problems, we model the obscuring elements as disks in the plane or spheres in space. In these cases, Y is the diameter of the object.

Let $G(y \mid \mathbf{x})$ denote the conditional c.d.f. of Y, given that the location of $\mathbf{p}$ is at $\mathbf{X} = \mathbf{x}$. $H(\mathbf{x})$ and $G(y \mid \mathbf{x})$ combined, yield the joint p.d.f. of $(\mathbf{X}, Y)$ on $\mathcal{S}$. Accordingly, if B is any measurable set of $(\mathbf{x}, y)$ values, the probability that a randomly chosen point of $\mathcal{P}$ has $(\mathbf{X}, Y)$ values in B is

$$P\{B\} = P\{(\mathbf{X}, Y)_i \in B\} = \iint_B h(\mathbf{x})g(y \mid \mathbf{x})d\mathbf{x}dy, \tag{1.47}$$

for all $i = 1, \cdots, N$; where $h(\mathbf{x})$ is the p.d.f. of $\mathbf{X}$ and $g(y \mid \mathbf{x})$ is the conditional p.d.f. of Y, given $\mathbf{X} = \mathbf{x}$.

Let $N\{B\}$ be the number of points in $\mathcal{P}$ whose $(\mathbf{X}, Y)$ values belong to B. $N\{B\}$ is a random variable. The collection $[N\{B\}; B \in \mathcal{B}_{\mathcal{S}} \times \mathcal{B}_Y]$ is called a **point process.** $\mathcal{B}_{\mathcal{S}}$ is the Borel σ-field* of $\mathbb{R}^n$ restricted to $\mathcal{S}$, and $\mathcal{B}_Y$ the σ-field generated by Y. Notice that $N = N\{\mathcal{S}\}$.

---

*All sets which we consider are such that their volume is given by $|S| = \int \underset{S}{\cdots} \int dx_1 \cdots dx_n$. Such sets are called "Borel measurable" sets.

*This is a technical term, pertaining to the collection of all Borel measurable sets, their complements and their unions. For a rigorous definition see any advanced probability text.

Given that $N = n$, the conditional distribution of $N\{B\}$ is the binomial $B(n, P\{B\})$. Let $\mathcal{Y}$ denote the range of Y (the marker) and $\Omega = \mathcal{S} \times \mathcal{Y}$. If $\{B_1, \cdots, B_m\}$, $m \geq 2$, is a measurable partition of Ω, the conditional distribution of $(N\{B_1\}, \cdots, N\{B_{m-1}\})$, given $N = n$, is m-nomial, with parameters n and $\boldsymbol{\theta} = (\theta_1, \cdots, \theta_{m-1})$ where $\theta_j = P\{B_j\}$, $j = 1, \cdots, m-1$. As we have previously shown, if N is a random variable having a Poisson distribution with mean λ (the intensity of the point process), the random variables $N\{B_j\}$, $j = 1, \cdots, m$, corresponding to any partition of Ω, are **independent** Poisson with means $\mu\{B_j\} = \lambda P\{B_j\}$, $j = 1, \cdots, m$. In this case the point process generates a **Poisson random field**.

A Poisson random field is called **homogeneous** on $\mathcal{S}$, if $h(\mathbf{x}) = c_{\mathcal{S}} d\mathbf{x}$, where $c_{\mathcal{S}}^{-1} = \int_{\mathcal{S}} \cdots \int d\mathbf{x}$. If, in addition, the distribution of Y is independent of $\mathbf{x}$, i.e., $G(y \mid \mathbf{x}) = G(y)$ for *all* $\mathbf{x} \in \mathcal{S}$, we say that the Poisson random field is **standard**. If N is fixed, the random field is called **multinomial**. In the following examples we illustrate Poisson and multinomial random fields.

Example 1.3. Consider a point process of N random intervals having $p = 2$ coordinates. X_1 represents the location of the left-end point of a random interval, and X_2 represents its random length. It is assumed that X_1 is distributed like $N(0, \sigma^2)$, and X_2 has an exponential distribution with mean l, independently of X_1. Moreover, the number N of intervals in S has a Poisson distribution with mean λ. What is the distribution of the number of intervals, $N\{\xi\}$, which cover the point ξ on the line? The probability that a point ξ on the line is covered by a random interval is

$$\begin{aligned} P\{C_\xi\} &= P[X_1 < \xi < X_1 + X_2] \\ &= \frac{1}{\sqrt{2\pi}\,\sigma} \int_{-\infty}^{\xi} e^{-\frac{1}{2}\left(\frac{x}{\sigma}\right)^2} \cdot e^{-\frac{\xi - x}{l}} dx \\ &= e^{-\frac{\xi}{l}} \cdot \frac{1}{\sqrt{2\pi}\,\sigma} \int_{-\infty}^{\xi} e^{-\frac{1}{2}\left(\frac{x}{\sigma}\right)^2 + \frac{x}{l}} dx \\ &= e^{-\frac{\xi}{l} + \frac{1}{2}\left(\frac{\sigma}{l}\right)^2} \Phi\left(\frac{\xi}{\sigma} - \frac{\sigma}{l}\right). \end{aligned} \tag{1.48}$$

Accordingly, $N\{\xi\}$ has a Poisson distribution with mean

$$\mu(\xi) = \lambda \cdot e^{-\frac{\xi}{l} + \frac{1}{2}\left(\frac{\sigma}{l}\right)^2} \Phi\left(\frac{\xi}{\sigma} - \frac{\sigma}{l}\right). \tag{1.49}$$

Notice that the probability that the point ξ remains uncovered is $e^{-\mu(\xi)}$. ■

Example 1.4. In the present example we develop a model for assessing the expected damage due to artillery shelling (see also Hall, 1988, pp. 32). Consider a region T in the plane, which is called the target. N shells are aimed at the center of T. Due to ballistic errors, the burst points of the shells is a point process. We assume that the coordinates of these burst points have bivariate normal distribution located at the center of T (the origin) and with dispersion parameters σ_1^2, σ_2^2 and coefficient of correlation $-1 < \rho < 1$. There are different models, which assess the damage. Let $d(|\boldsymbol{\xi} - \mathbf{X}|)$

be a damage function, which gives the amount of damage at the point $\boldsymbol{\xi} = (\xi_1, \xi_2)$, when a shell bursts at $\mathbf{X} = (X_1, X_2)$. We will consider here the "cookie-cutter" damage function which is

$$d(|\boldsymbol{\xi} - \mathbf{X}|) = \begin{cases} 1, & \text{if } |\boldsymbol{\xi} - \mathbf{X}| < R \\ 0, & \text{otherwise.} \end{cases}$$

In other words, all points $\boldsymbol{\xi}$ which are within distance R from the burst-point, are completely destroyed.

A point $\boldsymbol{\xi}$ "survives" a salvo of N shells, if the minimal distance from the N burst-points to $\boldsymbol{\xi}$ is greater than R. Let $I\{A\}$ denote an indicator variable, which assumes the value 1 if A is true and the value 0 if A is false. The survival function of a point $\boldsymbol{\xi}$, given $\mathbf{x}_1, \cdots, \mathbf{x}_N$, is

$$\delta(\boldsymbol{\xi}; \mathbf{x}_1, \cdots, \mathbf{x}_n) = \prod_{i=1}^{N} I\{|\boldsymbol{\xi} - \mathbf{X}_i| > R\}.$$

The fraction of T that survives (the proportion of the total vacancy of T to the area of T) is

$$F = \iint_T \delta(\xi_1 \xi_2; \mathbf{x}_1, \cdots, \mathbf{x}_N) d\xi_1 d\xi_2 / \iint_T d\xi_1 d\xi_2.$$

F is a random measure, since its value depends on the random realization of $\mathbf{X}_1, \cdots, \mathbf{X}_N$. The expected value of F is

$$E\{F\} = \iint_T \psi_N(\xi_1, \xi_2) d\xi_1 d\xi_2 / \iint_T d\xi_1 d\xi_2,$$

where

$$\psi_N(\xi_1, \xi_2) = (P[(X_1 - \xi_1)^2 + (X_2 - \xi_2)^2 > R^2])^N.$$

According to the above assumption about the distribution of (X_1, X_2), the probability that the point $\boldsymbol{\xi} = (\xi_1, \xi_2)$ is destroyed is

$$\begin{aligned} d(\xi_1, \xi_2) &= P[(X_1 - \xi_1)^2 + (X_2 - \xi_2)^2 \le R^2] \\ &= \frac{1}{\sqrt{2\pi}\,\sigma_1} \int_{\xi_1 - R}^{\xi_1 + R} e^{-\frac{1}{2\sigma_1^2}x^2} \left[\Phi\left(\frac{\xi_2 + (R^2 - (x - \xi_1)^2)^{1/2} - \rho \frac{\sigma_2}{\sigma_1} x}{\sigma_2 \sqrt{1 - \rho^2}} \right) \right. \\ &\quad \left. - \Phi\left(\frac{\xi_2 - (R^2 - (x - \xi_1)^2)^{1/2} - \rho \frac{\sigma_2}{\sigma_1} x}{\sigma_2 \sqrt{1 - \rho^2}} \right) \right] dx. \end{aligned} \tag{1.50}$$

Finally, $\psi_N(\xi_1, \xi_2) = (1 - d(\xi_1, \xi_2))^N$. Substituting this expression in the above equation yields a formula for $E\{F\}$. One has to integrate the above function numerically.

2

Geometrical Probability, Coverage and Visibility In Random Fields

The present chapter introduces some classical problems of geometrical probability coverage theory and random fields. The emphasis is, however, on practical aspects of the theory. The examples, which illustrate the general theory, focus on computational aspects. The objective in this chapter is to introduce basic concepts and techniques, which are encountered in later chapters.

2.1. Intersection of lines by random line segments

Consider the following classical problem (1777) which is known as the **Bouffon Needle Problem** (see Solomon (1978, p. 3)). A line segment, L, of length l is thrown at random on the plane. What is the probability that it intersects a given line $\mathcal{L}$ on the plane?

Without loss of generality, we can assume that $\mathcal{L}$ is the y-axis, in a rectangular coordinate system. Let (X_0, Y_0) be the coordinate of the random point at which the center of L rests. Obviously, L does not intersect $\mathcal{L}$ if $|X_0| > \frac{l}{2}$.

Let $\boldsymbol{\theta}$ be the random orientation angle of L. $\boldsymbol{\theta}$ is the angle between L and a line parallel to the x-axis, through (X_0, Y_0) (see Figure 2.1). We assume that X_0 has a uniform distribution on $(-\frac{l}{2}, \frac{l}{2})$, and that $\boldsymbol{\theta}$ has a uniform distribution on $(-\frac{\pi}{2}, \frac{\pi}{2})$.

L intersects $\mathcal{L}$ if, and only if, $\frac{|X_0|}{\cos(\boldsymbol{\theta})} < \frac{l}{2}$. Thus,

$$\begin{aligned} \Pr\{L \text{ intersects } \mathcal{L}\} &= \Pr\left\{-\frac{l}{2}\cos(\boldsymbol{\theta}) < X_0 < \frac{l}{2}\cos(\boldsymbol{\theta})\right\} \\ &= \frac{1}{l}\cdot\frac{1}{\pi}\int_{-\frac{\pi}{2}}^{\frac{\pi}{2}}\int_{-\frac{l}{2}\cos(\boldsymbol{\theta})}^{\frac{l}{2}\cos(\boldsymbol{\theta})} dx\,d\theta = \frac{1}{\pi}\int_{-\frac{\pi}{2}}^{\frac{\pi}{2}}\cos(\boldsymbol{\theta})d\theta \qquad (2.1) \\ &= \frac{2}{\pi} \end{aligned}$$

In the classical Bouffon's problem we consider parallel lines on the plane, with dis-ance d between any two lines. The problem is to determine the probability that L ntersects any of these lines, when thrown at random on the plane. In this case, we can ssume that $|X_0|$ is the distance between the center of L and the line closest to it. $|X_0|$ as a uniform distribution of $(0, \frac{d}{2})$, and the intersection probability is, according to .1), $\frac{2}{\pi}\cdot\frac{l}{d}$.

2. Random Lines Intersecting Circles

ippose that C is a circle of radius R, centered at the origin. Let $\mathcal{L}$ be a random line the plane. Let (A, B) be, respectively, the random intercept and random slope, of

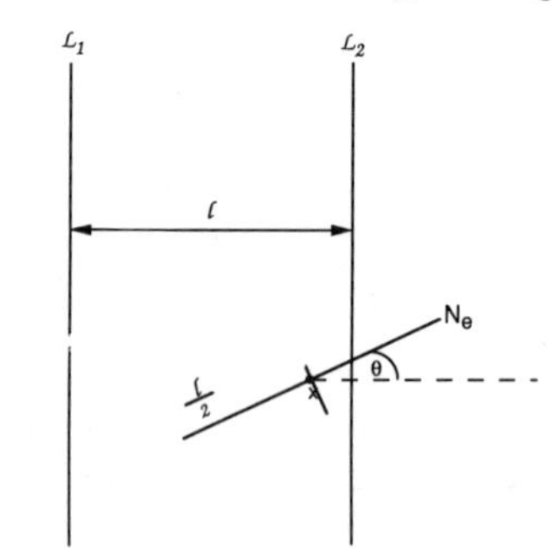

Figure 2.1. A Random Cord Intersecting a Line

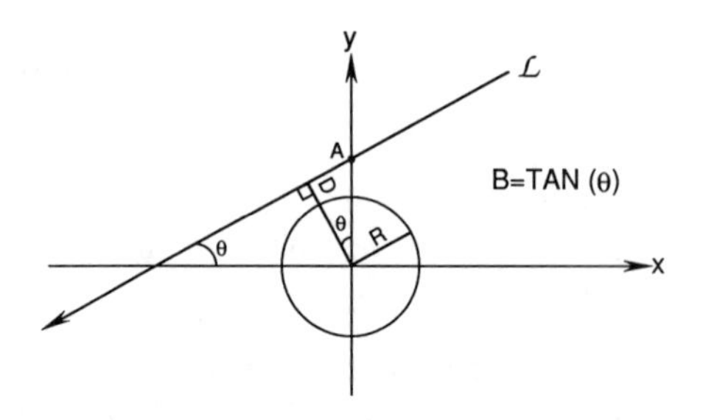

Figure 2.2. Random Line and A Circle

$\mathcal{L}$. Suppose that (A, B) have a joint distribution with a joint p.d.f. $f(a, b)$. What is the probability that $\mathcal{L}$ intersects $\mathcal{C}$? The distance of $\mathcal{L}$ from the origin $(0,0)$ is (see Figure 2.2) $D = |A|\cos(\tan^{-1}(B)) = \frac{|A|}{\sqrt{1+B^2}}$. $\mathcal{L}$ intersects $\mathcal{C}$ if, and only if, $D < R$. Hence,

$$\begin{aligned}\Pr\{\mathcal{L} \text{ intersects } \mathcal{C}\} &= \Pr\{\frac{|A|}{\sqrt{1+B^2}} < R\} \\ &= \int f_B(b)[F_{A|B}(R\sqrt{1+b^2} \mid b) - F_{A|B}(-R\sqrt{1+b^2} \mid b))]db \end{aligned} \tag{2.2}$$

where $f_B(b)$ is the marginal p.d.f. of the slope B, and $F_{A|B}(\cdot \mid b)$ is the conditional c.d.f of A, given $\{B = b\}$.

Example 2.1. Suppose that A and B are independent, uniformly distributed over $(-1, 1)$, and $R < 1$. We obtain from Eq. (2.2) that

$$\Pr\{\mathcal{L} \text{ intersects } \mathcal{C}\} = \frac{1}{2}\int_{-1}^{1} \min(1, R\sqrt{1+x^2})dx.$$

Since B has a uniform distribution on $(-1, 1)$, the p.d.f. of B^2 is

$$f_{B^2}(x) = \frac{1}{2}x^{-1/2}, \quad 0 \le x \le 1.$$

Thus,

$$\Pr\{\mathcal{L} \text{ intersects } \mathcal{C}\} = \frac{1}{2}\int_0^1 \min(1, R\sqrt{1+y})y^{-1/2}dy$$

$$= \begin{cases} \frac{R}{2}\int_0^1 (1+y)^{1/2}y^{-1/2}dy, & \text{if } R \le \frac{1}{\sqrt{2}} \\ \frac{R}{2}\int_0^{(1-R^2)/R^2}(1+y)^{1/2}y^{-1/2}dy + \frac{1}{2}\int_{(1-R^2)/R^2}^1 y^{-1/2}dy, & \text{if } \frac{1}{\sqrt{2}} < R \le 1 \end{cases} \quad (2.3)$$

In the following table we present some of these intersection probabilities as a function of R. The integration in (2.3) was performed numerically, using the software MATHCAD®.

R	0.1	0.2	0.3	0.4	0.5	0.6	0.7
Pr	0.1148	0.2296	0.3443	0.4591	0.5739	0.6887	0.8036

R	0.707	0.816	0.913	0.953	1.000
Pr	0.8116	0.915	0.965	0.990	1.000

■

We consider now the random size (length) of the cord created by the intersection of a random line $\mathcal{L}$ with the circle $\mathcal{C}$.

We have seen that the condition for $\mathcal{L}$ to intersect $\mathcal{C}$ is that $|A|/(1+B^2)^{1/2} < R$. Under this condition, the length D of the intersecting cord is

$$D = 2\left(R^2 - \frac{A^2}{1+B^2}\right)^{1/2}.$$

Generally, the length of the intersecting cord of a random line $\mathcal{L}$ with a circle $\mathcal{C}$, centered at the origin, is

$$D = \begin{cases} 0, & \text{if } R \le \frac{|A|}{(1+B^2)^{1/2}} \\ 2(R^2 - \frac{A^2}{1+B^2})^{1/2}, & \text{if } R > \frac{|A|}{(1+B^2)^{1/2}}. \end{cases} \quad (2.4)$$

Accordingly, the c.d.f. of D has a jump at $D = 0$ of size

$$p_0 = \Pr\{D = 0\} = 1 - \Pr\{\mathcal{L} \text{ intersects } \mathcal{C}\}. \quad (2.5)$$

Recall that $0 \le D \le 2R$. Thus, for $0 < d < 2R$,

$$\Pr\{D \le d\} = 1 - \int \left[F_{A|B}\left(\left(R^2 - \frac{d^2}{4}\right)^{1/2}(1+b^2)^{1/2}\right) - F_{A|B}\left(-\left(R^2 - \frac{d^2}{4}\right)^{1/2}(1+b^2)^{1/2}\right)\right] f_B(b)db. \quad (2.6)$$

Example 2.2. As in Example 2.1, let A and B be independent, having uniform distributions on $(-1,1)$. The expected length of the intersecting cord, when $R=1$, is

$$E\{D\} = \frac{1}{2}\int_{-1}^{1}\int_{-1}^{1}\left(1-\frac{x^2}{1+y^2}\right)^{1/2} dxdy = 1.68575 \tag{2.7}$$

This result obviously depends on the distribution of (A,B). ■

One can parametrize a line $\mathcal{L}$ in the plane by parameters (ρ,θ), where ρ is the distance of $\mathcal{L}$ from the origin $(0,0)$ and θ is the (counter clockwise) angle between the normal to $\mathcal{L}$ through the origin and the x-axis. In terms of these parameters

$$D = \left[\left(1-\left(\frac{\rho}{R}\right)^2\right)^+\right]^{1/2}, \tag{2.8}$$

where $(a)^+ = \max(0,a)$. The expected value of D, for a "uniform" distribution of (ρ,θ) is

$$\begin{aligned} E\{D\} &= 2\int_0^{2\pi}\int_0^R\left(1-\left(\frac{\rho}{R}\right)^2\right)^{1/2} d\rho d\theta = 4\pi\int_0^R\left(1-\left(\frac{\rho}{R}\right)^2\right)^{1/2} d\rho \\ &= 2\pi R^2\int_0^1 u^{-1/2}(1-u)^{1/2}du = 2\pi R^2 B\left(\frac{1}{2},\frac{3}{2}\right) \\ &= 2\pi R^2\cdot\frac{\Gamma(\frac{1}{2})\cdot\frac{1}{2}\Gamma(\frac{1}{2})}{\Gamma(2)} = \pi^2R^2. \end{aligned} \tag{2.9}$$

We have shown that, if $\mathcal{C}^*$ is the disk with circumference $\mathcal{C}$,

$$\int_0^{2\pi}\int_0^{\infty} \text{Length } (\mathcal{L}\cap\mathcal{C}^*)d\rho d\theta = \pi \text{ area}(\mathcal{C}^*). \tag{2.10}$$

This is a special case of a classical result in the theory of geometrical probabilities, in which $\mathcal{C}^*$ is any convex set in $\mathbb{R}^2$ (see Baddeley (1982), Solomon (1978)).

2.3. Random Disks Intersecting Lines

The intersection of lines of sight by randomly placed disks, in a region of the plane, is fundamental topic in the theory of stochastic visibility, as will be shown in the coming chapters. In the present section we provide some basic results.

A random disk D is specified by three random variables (X,Y,R), having a joint distribution $F(x,y,r)$. (X,Y) are the rectangular coordinates of its center, and R is the radius of the disk. The random variables X, Y and R may or may not be independent. If convenient, the random coordinates of the center of a disk can be presented in polar form (ρ,θ).

Let $\mathcal{L}$ be a specific (non-random) line in the plane, whose equation is $\mathcal{L}: y=\alpha+\beta x$, $-\infty<x<\infty$. A random disk D intersects $\mathcal{L}$ if, and only if, the distance of its center (X,Y) from $\mathcal{L}$ is smaller than its radius R.

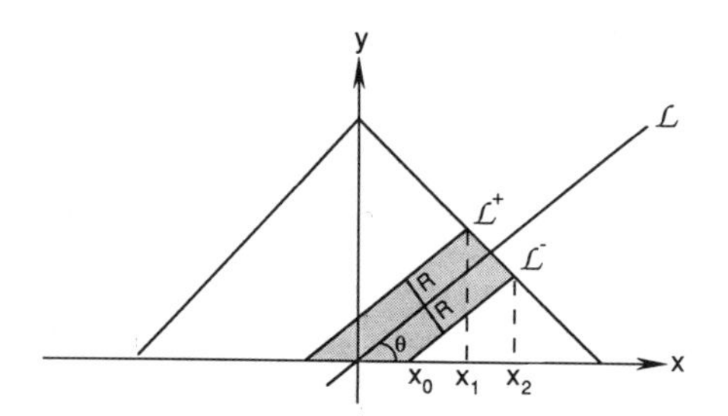

Figure 2.3. The Set of Disk Centers Intersecting $\mathcal{L}$

Let $H_{R|X,Y}(r \mid x, y)$ be the conditional c.d.f. of R, given $\{X = x, Y = y\}$ and let $f(x, y)$ be the joint p.d.f. of (X, Y). Since D intersects $\mathcal{L}$ only if $R > |Y - \alpha - \beta X|/(1 + \beta^2)^{1/2}$ (see Figure 2.2), the probability that D intersects $\mathcal{L}$ is

$$\Pr\{D \text{ intersects } \mathcal{L}\} = 1 - \int\int H_{R|X,Y}\left(\frac{|y - \alpha - \beta x|}{\sqrt{1+\beta^2}} \mid x, y\right) f(x, y) dx dy. \quad (2.11)$$

Example 2.3. We develop now Eq. (2.11) for the special case of R independent of (X, Y). Furthermore, (X, Y) have a uniform distribution over the triangle T, whose vertices are $(-1, 0)$, $(1, 0)$, and $(0, 1)$ (see Figure 2.4). The line $\mathcal{L}$ passes through the origin with slope $\beta = \tan(\theta)$, $0 < \theta < \frac{\pi}{2}$. A disk with radius R intersects $\mathcal{L}$ only if it is centered between the lines $\mathcal{L}^-$ and $\mathcal{L}^+$, which are parallel to $\mathcal{L}$ and of distance R from $\mathcal{L}$.

Notice that the area of T is 1. Thus, the conditional probability that a disk of radius R intersects $\mathcal{L}$, given $\{R = r\}$, is the area of the region in T between $\mathcal{L}^-$ and $\mathcal{L}^+$. The equations of $\mathcal{L}^-$ and $\mathcal{L}^+$ are,

$$\mathcal{L}^- : y = R\sqrt{1+\beta^2} + \beta x,$$

and

$$\mathcal{L}^+ : y = -R\sqrt{1+\beta^2} + \beta x.$$

To simplify, we assume that $\max\{\frac{R}{\beta}\sqrt{1+\beta^2}, R\sqrt{1+\beta^2}\} < 1$.

$\mathcal{L}^-$ intersects the x-axis at $x_0 = \frac{R}{\beta}\sqrt{1+\beta^2}$.

Let $\mathcal{L}^*$ be the side of T connecting $(1, 0)$ and $(0, 1)$.

Let x_1 be the point at which $\mathcal{L}^+$ intersects $\mathcal{L}^*$. $x_1 = (1 - R\sqrt{1+\beta^2})/(1 + \beta)$. The line $\mathcal{L}^-$ intersects $\mathcal{L}^*$ at (x_2, y_2), where $x_2 = (1 + R\sqrt{1+\beta^2})/(1 + \beta)$ and $y_2 = \frac{\beta}{1+\beta} - R\frac{\sqrt{1+\beta^2}}{1+\beta}$. In terms of these variables, the conditional probability that D

intersects $\mathcal{L}$, given R, is

$$\Pr\{D \text{ intersects } \mathcal{L} \mid R\} = \frac{1}{2} - \frac{1}{2}(1-x_0)y_2 - \frac{1}{2}(1-R\sqrt{1+\beta^2})x_1 + \frac{1}{2}\frac{R^2(1+\beta^2)}{\beta} = 2R\frac{\sqrt{1+\beta^2}}{1+\beta}. \tag{2.12}$$

Finally, from the law of the iterated expectation (see Problem [1.5.4]) we obtain

$$\Pr\{D \text{ intersects } \mathcal{L}\} = 2\frac{\sqrt{1+\beta^2}}{1+\beta}E\{R\}. \tag{2.13}$$

■

2.4. The Coverage of A Circle By Random Arcs

The following is a classical problem in the theory of geometric probabilities (see H. Solomon, 1978, pp. 76). Consider a circle $\mathcal{C}$ having a circumference of length $L = 1$. n (black) arcs of fixed length a, $a < 1$, are randomly placed on the circle $\mathcal{C}$. What is the probability that the n black arcs will cover $\mathcal{C}$? Obviously, if $n < [1/a]$* then the probability is zero that $\mathcal{C}$ is completely covered. We thus assume that the number of arcs, n, is greater than $k = [1/a]$.

Consider a clockwise orientation on $\mathcal{C}$ (see Figure 2.4). Each black arc A_i ($i = 1, \cdots, n$) on $\mathcal{C}$ has a left-end and a right-end point. We say that a black arc A is followed by a gap, if some points of $\mathcal{C}$ to the right of A are uncovered. A gap follows an arc A if the left-end point of any other arc is *not* within A. Thus, since the left-end points of the black arcs are randomly and independently put on $\mathcal{C}$, the probability of the event E_i that a gap follows arc A_i is

$$p_i = (1-a)^{n-1}, \quad i = 1, 2, \cdots, n. \tag{2.14}$$

Let E_{ij}, $1 \le i < j \le n$, denote the events that both black arcs A_i and A_j are followed by gaps. The probability of this event is

$$p_{ij} = (1-2a)^{n-1}, \quad 1 \le i < j \le n. \tag{2.15}$$

Indeed, there are gaps following both A_i *and* A_j if and only if the left-end points c $n-1$ black arcs are distributed on $\mathcal{C} - (A_i \cup A_j)$ which has an arc length of $(1-2a$ Similarly, for any $1 \le l \le k$, $1 \le j_1 < j_2 < \cdots < j_l \le n$, let $E_{j_1 j_2 \cdots j_l}$ denote the eve that arcs $A_{j_1}, \cdots, A_{j_l}$ are followed by gaps. Let $p_{j_1 \cdots j_l}$ denote the probability of $E_{j_1 \cdots}$ Argument like above yields

$$p_{j_1 \cdots j_l} = (1-la)^{n-1}. \tag{2.16}$$

*$[x]$ is the largest integer which is not greater than x.

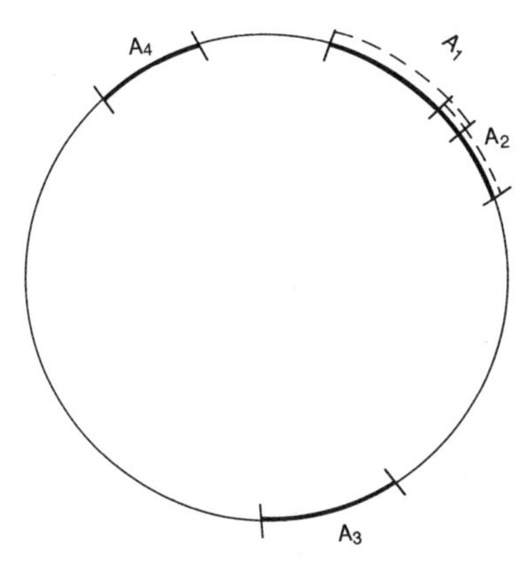

Figure 2.4. Random Arcs On A Circle

Let $S_l = \sum_{1 \le j_1 < \cdots < j_l \le n} p_{j_1 \cdots j_l} = \binom{n}{l}(1 - la)^{n-1}$, $l = 1, 2, \cdots, k$. Notice that $p_{j_1 \cdots j_l} = 0$ f $l > k$. The probability of at least one gap is (see Feller, 1968, p. 98)

$$\Pr\{E_1 \cup E_2 \cup \cdots \cup E_n\} = \sum_{l=1}^{k} (-1)^{l-1} S_l. \tag{2.17}$$

inally, since complete coverage of $\mathcal{C}$ is the complement of the event of at least one gap, he formula for the probability of the coverage of $\mathcal{C}$ is

$$\begin{aligned} \Pr\{\text{Coverage}\} &= 1 - \sum_{l=1}^{k} (-1)^{l-1} S_l \\ &= \sum_{l=0}^{k} (-1)^l \binom{n}{l} (1 - la)^{n-1} \end{aligned} \tag{2.18}$$

his formula was first derived by Stevens (1939). ■

The problem of covering the circle by n arcs of **random** length, placed on $\mathcal{C}$ at ndom, was studied by various authors. In particular, see Hüsler (1982), Siegel (1978, 79), and Siegel and Holst (1982).

5. Vacancies On The Circle

before, let $\mathcal{C}$ be a circle centered at the origin, with radius $R = 1$. We throw on $\mathcal{C}$ a ndom arc **A**. A point **P** on $\mathcal{C}$ which is not covered by **A** is called **vacant**. An arc from to $\mathbf{P}_2$ on $\mathcal{C}$ is called **completely vacant** if all its points are vacant. The random rdinates of a random arc **A** are its right-end limit X and its length Y. We assume t X and Y are **independent**; X has a uniform distribution on $(0, 2\pi)$ and Y has a f. $F_Y(y)$ on the interval, $0 \le y \le \pi$; i.e., $F_Y(y) = 0$ for all $y < 0$ and $F_Y(y) = 1$ for $y > \pi$.

Let $q(t)$ denote the probability that the arc from $(1,0)$ to $(1,t)$, $0 < t \leq \pi$, will be completely vacant. This probability is given by

$$q(t) = 1 - \frac{1}{2\pi}(t + E\{Y\} - \psi(2\pi - t)), \tag{2.19}$$

where

$$\psi(t) = \begin{cases} \int_t^{\pi}(1 - F_Y(y))dy, & 0 < t < \pi. \\ \\ 0, & \pi \leq t < 2\pi. \end{cases} \tag{2.20}$$

The derivation of Eq. (2.19) is given as Problem [2.5.1].

Let $\mathbf{P}_1, \cdots, \mathbf{P}_n$ $(n \geq 2)$ be n points on $\mathcal{C}$, specified by their polar coordinates, $(1, s_i)$, $i = 1, \cdots, n$, $0 < s_1 < s_2 < \cdots < s_n < 2\pi$. Let $t_0 = 2\pi - s_n + s_1$ and $t_i = s_{i+1} - s_i$, $i = 1, \cdots, n-1$. t_i is the length of the arc on $\mathcal{C}$, between $\mathbf{P}_{s_i - 1}$ and $\mathbf{P}_{s_i}$. We derive now the probability that all the n points are **simultaneously vacant**.

Define B_0 to be the event that the random arc $\mathbf{A}$ lies entirely between $\mathbf{P}_{s_n}$ and $\mathbf{P}_{s_1}$, and let B_i $(i = 1, \cdots, n-1)$ be the event that $\mathbf{A}$ lies between $\mathbf{P}_{s_i}$ and $\mathbf{P}_{s_{i+1}}$. $B_0, \cdots, B_{n-1}$ are disjoint events. The probability that neither one of the n points is covered by $\mathbf{A}$ (simultaneous vacancy) is

$$\begin{aligned} p(t_1, \cdots, t_{n-1}) &= \sum_{i=0}^{n-1} \Pr\{B_i\} \\ &= \sum_{i=0}^{n-1} q(2\pi - t_i) \\ &= 1 - \frac{n}{2\pi}E\{Y\} + \frac{1}{2\pi}\sum_{i=0}^{n-1}\psi(t_i). \end{aligned} \tag{2.21}$$

2.6. Vacancies On The Plane

Let D be a random disk in the plane, having random coordinates (ρ, θ, R), where (ρ, θ) are the polar coordinates of its center, and R is its radius. A point $\mathbf{P}$ in the plane said to be **vacant**, if the random disk, D, does not cover $\mathbf{P}$. A point which is not vacant is said to be **covered**.

A set of points S is said to be **completely vacant** if **all** points of S are vacant. Let $h(\rho, \theta)$ be the p.d.f. of (ρ, θ) and $G(r \mid \rho, \theta)$ the conditional c.d.f. of R, given (ρ, θ). Thus, the probability that a point $\mathbf{P}_0 = (\rho_0, \theta_0)$ is vacant is

$$\psi(\rho_0, \theta_0) = P\{(\rho^2 + \rho_0^2 - 2\rho\rho\cos(\theta_0 - \theta))^{1/2} > R\}. \tag{2.22}$$

Indeed, the distance between the center of D and $\mathbf{P}_0$ should be greater than R. Thus

$$\psi(\rho_0, \theta_0) = \int_0^{\infty}\int_0^{2\pi} h(\rho, \theta) \cdot G((\rho^2 + \rho_0^2 - 2\rho\rho_0\cos(\theta - \theta_0))^{1/2} \mid \rho, \theta)d\rho d\theta. \tag{2.23}$$

Let

$$I(\rho_0, \theta_0) = \begin{cases} 1, & \text{if } \mathbf{P} = (\rho_0, \theta_0) \text{ is vacant} \\ 0, & \text{otherwise.} \end{cases}$$

A **measure of total vacancy** of a set S is (see Hall (1988), p. 21),

$$\text{vac}\{S\} = \iint_{(\rho_0,\theta_0)\in S} I(\rho_0, \theta_0) d\rho_0 d\theta_0. \tag{2.24}$$

Since $E\{I(\rho_0, \theta_0)\} = \psi(\rho_0, \theta_0)$, the expected value of vac$\{S\}$ is

$$E\{\text{vac}\{S\}\} = \iint_{(\rho_0,\theta_0)\in S} \psi(\rho_0, \theta_0) d\rho_0 d\theta_0. \tag{2.25}$$

This formula is called the **Robbins Formula** (see H. Robbins (1944, 45)).

2.6.1. The Vacancy of a Point In The Plane Under Bivariate Normal Distribution of Disk Centers

We develop here formula (2.23) for the special case where disk centers have a bivariate normal distribution. The random disk radius, R, is independent of the disk center. In the present case it is more convenient to use rectangular coordinates. Without loss of generality, assume that (X, Y) has a bivariate normal distribution centered at the origin $(0, 0)$, with covariance matrix

$$\Sigma = \begin{pmatrix} \sigma_x^2 & \rho\sigma_x\sigma_y \\ \rho\sigma_x\sigma_y & \sigma_y^2 \end{pmatrix}, \quad 0 < \sigma_x^2, \ \sigma_y^2 < \infty, \quad -1 < \rho < 1.$$

We assume that the disk radius, R, is independent of (X, Y), and has a uniform distri-bution on (a, b).

Let $\mathbf{P} = (x_0, y_0)$. The probability that $\mathbf{P}$ is covered is

$$\begin{aligned} &\Pr\{(X - x_0)^2 + (Y - y_0)^2 \leq R^2\} \\ &= \frac{1}{b-a} \int_a^b \cdot \frac{1}{\sigma_y} \int_{y_0-r}^{y_0+r} \phi\left(\frac{y}{\sigma_y}\right) \left[\Phi\left(\frac{x_0 + (r^2 - (y - y_0)^2)^{1/2} - \beta y}{\sigma_x\sqrt{1-\rho^2}}\right)\right. \\ &\left. - \Phi\left(\frac{x_0 - (r^2 - y - y_0)^2)^{1/2} - \beta y}{\sigma_x\sqrt{1-\rho^2}}\right)\right] dy dr, \end{aligned} \tag{2.26}$$

where $\beta = \rho\frac{\sigma_x}{\sigma_y}$. In developing Eq. (2.26) we have used the results of Problem [1.7.10] and the fact that (X, Y) is independent of R. This coverage probability, as given by Eq. (2.26), can be easily determined numerically. A general analytic expression for (2.26) is quite complicated (see Eckler and Burr (1972) for various formulae of circular coverage probability). We show here an analytic derivation for the special case of $\sigma_x^2 = \sigma_y^2 = 1$ and $\rho = 0$. In this particular case $(X - x_0)^2 + (Y - y_0)^2$ is distributed like the non-central chi-square $\chi^2[2; \mu]$, where the parameter of non-centrality is $\mu = \frac{1}{2}(x_0^2 + y_0^2)$.

Furthermore, as shown in Problem [1.7.18], $\chi^2[2;\mu]$ is distributed like a Poisson mixture of central chi-squares $\chi^2[2+2j]$. Thus, since R is independent of (X,Y),

$$\begin{aligned}\Pr\{(X-x_0)^2+(Y-y_0)^2>r^2 \mid R=r\} \\ &= \Pr\{\chi^2[2;\mu]>r^2\} \\ &= e^{-\mu}\sum_{j=0}^{\infty}\frac{\mu^j}{j!}\Pr\{\chi^2[2+2j]>r^2\} \\ &= e^{-\mu}\sum_{j=0}^{\infty}\frac{\mu j}{j!}P\left(j;\frac{r^2}{2}\right),\end{aligned} \tag{2.27}$$

where $P(j;\frac{r^2}{2}) = e^{-r^2/2}\sum_{i=0}^{j}\frac{(r^2/2)^i}{i!}$ is the c.d.f. of a Poisson distribution with mean $\frac{r^2}{2}$. By changing the order of summation we can write,

$$\begin{aligned}\Pr\{\chi^2[2;\mu]>r^2\} &= e^{-r^2/2}\sum_{j=0}^{\infty}\frac{r^{2j}}{2^j j!}(1-P(j-1;\mu)) \\ &= 1-e^{-r^2/2}\sum_{j=1}^{\infty}\frac{r^{2j}}{2^j j!}P(j-1;\mu).\end{aligned} \tag{2.28}$$

Finally, the vacancy probability of $\mathbf{P}_0$ is

$$\begin{aligned}\Pr\{(X-x_0)^2+(Y-y_0)^2>R^2\} &= E\{\Pr\{(X-x_0)^2+(Y-y_0)^2>R^2 \mid R\}\} \\ &= 1-\sum_{j=1}^{\infty}\frac{1}{2^j j!}P(j-1;\mu)E\{e^{-R^2/2}R^{2j}\}.\end{aligned} \tag{2.29}$$

2.6.2. Complete Vacancy Of Triangles

In the present section we also assume that the random coordinates of disk centers (X,Y), have a bivariate normal distribution, centered at the origin, and that the random radius, R, is independent of (X,Y).

Let $\Delta(x_0,y_0,x_1,y_1,x_2,y_2)$ be a triangle with vertices $\mathbf{P}_0=(x_0,y_0)$, $\mathbf{P}_1=(x_1,y_1)$ and $\mathbf{P}_2=(x_2,y_2)$. We label the vertices so that $y_0=\min\{y_0,y_1,y_2\}$. $\mathbf{P}_1$ is the adjacent vertex, when we move along the sides of Δ in a counter clockwise direction.

Given $\{R=r\}$, Δ is completely vacant if the disk center (X,Y) is outside the extended triangle Δ_r, where the sides of Δ_r are parallel to those of Δ and are a distance r from them. Moreover, $\Delta\subset\Delta_r$ (see Figure 2.5).

Let $\mathbf{P}_i^+=(x_i^+,y_i^+)$, $i=0,1,2$, be the vertices of Δ_r. The probability that Δ completely vacant is

$$\psi(\Delta)=1-E\{\Pr(X,Y)\in\Delta_R \mid R\}\}. \tag{2.30}$$

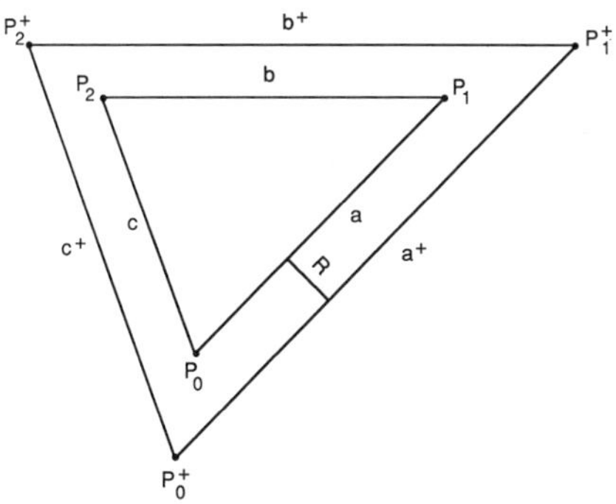

Figure 2.5. The Geometry of Δ and Δ_r

We develop now the formula for

$$H_r(\Delta) = \Pr\{(X,Y) \in \Delta_r \mid R = r\}. \tag{2.31}$$

Make the orthogonal transformation

$$\begin{pmatrix} X' \\ Y' \end{pmatrix} = A \begin{pmatrix} X \\ Y \end{pmatrix}$$

so that the side $\overline{P_1 P_2}$ of Δ, or $\overline{P_1^+ P_2^+}$ of Δ_r, are parallel to the x-axis, and $y_0' = \min\limits_{0 \le i \le 2} (y_i')$. The orthogonal matrix, A, applied for this purpose is

$$A = \frac{1}{(1+B^2)^{1/2}} \begin{bmatrix} 1 & B \\ -B & 1 \end{bmatrix}, \quad \text{if } x_1 \neq x_2$$

where $B = (y_2 - y_1)/(x_2 - x_1)$. If $x_1 = x_2$ then

$$A = \begin{bmatrix} 0 & -1 \\ 1 & 0 \end{bmatrix}, \quad \text{if } x_0 < x_1$$

nd

$$A = \begin{bmatrix} 0 & -1 \\ -1 & 0 \end{bmatrix}, \quad \text{if } x_0 > x_1.$$

ccording to Problem [1.7.13], the distribution of $\begin{pmatrix} X' \\ Y' \end{pmatrix}$ is bivariate normal with mean ,0) and covariance matrix

$$\Sigma = A\,\Sigma A' = \begin{bmatrix} (\sigma_x')^2 & \rho'\sigma_x'\sigma_y' \\ \rho'\sigma_x'\sigma_y' & (\sigma_y')^2 \end{bmatrix}.$$

e triangles Δ and Δ_r are transformed to Δ' and Δ_r', where the vertices of Δ' and . are $\mathbf{P}_i = (x_i', y_i')$ and $\mathbf{P}_i^+ = (x_i^+, y_i^+)$, respectively,

$$\begin{pmatrix} x_i' \\ y_i' \end{pmatrix} = A \begin{pmatrix} x_i \\ y_i \end{pmatrix}, \quad i = 0, 1, 2.$$

Thus, since (X, Y) are independent of R, the conditional probability that $(X, Y) \in \Delta_R$, given $\{R = r\}$ is $H_r(\Delta) = \Pr\{(X', Y') \in \Delta_r'\}$.

Let $y' = a_1(r) + b_1 x'$ be the equation of the line passing through $\mathbf{P}_0^{+\prime}$ and $\mathbf{P}_1^{+\prime}$. Similarly, let $y' = a_2(r) + b_2 x'$ be the equation of the line passing through $P_0^{+\prime}$ and $P_2^{+\prime}$. Assuming that these two lines are not parallel to the y-axis, we obtain that $b_1 = \dfrac{y_0' - y_0'}{x_1' - x_0'}$ and $b_2 = \dfrac{y_1' - y_0'}{x_2' - x_0'}$ since $y_1' = y_2'$. Furthermore,

$$a_1(r) = (y_0' - b_1 x_0') - r\sqrt{1 + b_1^2}$$

and

$$a_2(r) = (y_0' - b_2 x_0') - r\sqrt{1 + b_2^2}.$$

The coordinates of the vertices of Δ_r' are:

$$\begin{aligned}
x_0^+(r) &= \frac{a_2(r) - a_1(r)}{b_1 - b_2} \\
y_0^+(r) &= a_1(r) + b_1 x_0^+(r) \\
x_1^+(r) &= \frac{y_1' - y_0'}{b_1} - x_0' - r\left(1 + \frac{1}{b_1^2}\right)^{1/2} \\
y^+(r) &= y_1' + r \\
x_2^+(r) &= \frac{y_1' - y_0'}{b_2} - x_0' - r\left(1 + \frac{1}{b_2^2}\right)^{1/2} \\
y_2^+(r) &= y_1' + r.
\end{aligned}$$

Assuming that $x_2^+(r) < x_0^+(r) < x_1^+(r)$, we obtain

$$\begin{aligned}
H_r(\Delta) &= \Pr\{x_2^+(r) < X' < x_0^+(r), a_2(r) + b_2 X' < Y' < y_1' + r \mid R = r\} \\
&\quad + \Pr\{x_0^+(r) < X' < x_1^+(r), a_1(r) + b_1 X' < Y' < y_1' + r \mid R = r\} \\
&= \frac{1}{\sigma_x'} \int_{x_2^+(r)}^{x_1^+(r)} \phi\left(\frac{x}{\sigma_x'}\right) \Phi\left(\frac{y_1' + r - \frac{\sigma_y'}{\sigma_x'}\rho' x}{\sigma_y'\sqrt{1 - (\rho')^2}}\right) dx \\
&\quad - \frac{1}{\sigma_x'} \int_{x_2^+(r)}^{x_0^+(r)} \phi\left(\frac{x}{\sigma_x'}\right) \Phi\left(\frac{a_2(r) + b_2 x - \frac{\sigma_y'}{\sigma_x'}\rho' x}{\sigma_y'\sqrt{1 - (\rho')^2}}\right) dx \\
&\quad - \frac{1}{\sigma x'} \int_{x_0^+(r)}^{x_1^+(r)} \phi\left(\frac{x}{\sigma_x'}\right) \Phi\left(\frac{a_1(r) + b_1(x) - \frac{\sigma_y'}{\sigma_x'}\rho_x'}{\sigma_y'\sqrt{1 - (\rho')^2}}\right) dx.
\end{aligned} \tag{2.3}$$

Finally, the expected value of $H_R(\Delta)$ has to be evaluated. When R has a unifor distribution on $(0, b)$ then

$$E\{H_R(\Delta)\} = \frac{1}{b}\int_0^b H_r(\Delta)\, dr. \tag{2.3}$$

Also this integral can be evaluated numerically. Numerical examples are given in Problems [2.6.4] and [2.6.5].

2.7. Visibility Of Points On A Circle In A Poisson Random Field

In the present section we consider a circle $\mathcal{C}$ of radius $R = 1$, centered at the origin. N disks are randomly placed within $\mathcal{C}$. We assume that these random disks do not cover the origin, and do not intersect $\mathcal{C}$. Let (ρ, θ) be the polar coordinates of the disk center, and let Y be the diameter of a random disk, $0 < \rho < 1$ and $0 \le \theta < 2\pi$. Moreover, $Y/2 < \min(\rho, 1-\rho)$. Thus, the sample space of (ρ, θ, y) is

$$\Omega = \{(\rho, \theta, y) : \frac{1}{2}y < \rho < 1 - \frac{1}{2}y;\ 0 < \theta < 2\pi,\ 0 < y < 1\}.$$

We further assume that θ has a uniform distribution on $(0, 2\pi)$; Y has a distribution with density $g(y)$ and the conditional p.d.f. of ρ, given $\{Y = y\}$, is $h(\rho \mid y) = \dfrac{2\rho}{1-y}$, $\dfrac{y}{2} < \rho < 1 - \dfrac{y}{2}$. Thus, if N has a Poisson distribution with mean μ per unit area,

$$E\{N\{\Omega\}\} = 2\mu \int_0^1 \int_0^{2\pi} \int_{y/2}^{1-y/2} \frac{\rho}{1-y} g(y) d\rho d\theta dy = \pi\mu. \tag{2.34}$$

Suppose that there is a source of light and an observer in the origin. A disk within $\mathcal{C}$ casts a shadow on $\mathcal{C}$, which is a **shadow arc** of size

$$X(\rho, \theta, Y) = 2\sin^{-1}\left(\frac{Y}{2\rho}\right). \tag{2.35}$$

Given $N\{\Omega\} = n$ random disks within $\mathcal{C}$, there are n random shadow arcs cast on $\mathcal{C}$ by the random disks. The lengths of these shadow arcs, $X_1, \cdots, X_n$, are conditionally, given $N\{\Omega\} = n$, independent random variables, having a common c.d.f.

$$F(x) = 2\int_0^{D(x)} \frac{1}{1-y}\left(\int_{\frac{y}{\sin(\frac{x}{2})}}^{1-y/2} \rho d\rho\right) g(y) dy, \tag{2.36}$$

where

$$D(x) = \frac{2\sin(\frac{x}{2})}{1+\sin(\frac{x}{2})} = 1 - \tan^2\left(\frac{\pi - x}{4}\right). \tag{2.37}$$

Notice that

$$2\int_{\frac{y}{2\sin(\frac{x}{2})}}^{1-y_2} \rho d\rho = 1 - y + \frac{y^2}{4}\cot^2\left(\frac{x}{2}\right). \tag{2.38}$$

Hence,

$$F(x) = \int_0^{D(x)} \left(1 - \frac{y^2}{4(1-y)}\cot^2\left(\frac{x}{2}\right)\right) g(y) dy. \tag{2.39}$$

Suppose that r, $r \ge 2$, points are specified on $\mathcal{C}$, and we are interested in the probability that all these r points are simultaneously visible. We can substitute the c.d.f. $F(x)$ into

Eq. (2.20) to obtain the formula for $q_1(t)$, with which we can evaluate $p(t_1, \cdots, t_{r-1})$, according to Eq. (2.21). Finally, the probability that all the r points are simultaneously visible is

$$\psi(t_1, \cdots, t_{r-1}) = e^{-\mu\pi} \sum_{n=0}^{\infty} \frac{(\mu\pi)^n}{n!} (p(t_1, \cdots, t_{r-1}))^n. \tag{2.40}$$

In the special case of $g(y) = 2(1-y)$, $0 < y < 1$, we obtain

$$\begin{aligned} F(x) &= 2\int_0^{D(x)} (1-y)dy - \frac{1}{2}\cot^2\left(\frac{x}{2}\right) \cdot \int_0^{D(x)} y^2 dy \\ &= 1 - \tan^4\left(\frac{\pi - x}{4}\right) - \frac{2}{3}\tan^2\left(\frac{\pi - x}{4}\right)\left(1 - \tan^2\left(\frac{\pi - x}{4}\right)\right), \end{aligned} \tag{2.41}$$

$0 \le x \le \pi$. The expected value of X is then

$$\begin{aligned} E\{X\} &= \int_0^{\pi} (1 - F(x))dx \\ &= \int_0^{\pi} \left[\tan^2\left(\frac{\pi - x}{4}\right) + \frac{2}{3}\tan^2\left(\frac{\pi - x}{4}\right)\left(1 - \tan^2\left(\frac{\pi - x}{4}\right)\right)\right] dx. \end{aligned} \tag{2.42}$$

Making the change of variable $y = \tan\left(\frac{\pi - x}{4}\right)$ we obtain

$$\begin{aligned} E\{X\} &= 4\int_0^1 \frac{y^2}{1+y^2}dy + \frac{8}{3}\int_0^1 \frac{y^2(1-y^2)}{1+y^2}dy \\ &= 4 - \pi + \frac{8}{3}\left(2 - \frac{1}{3} - \frac{\pi}{2}\right) = \frac{76 - 21\pi}{9} \\ &= 1.1141. \end{aligned}$$

The function $\phi(t)$ assumes the form

$$\begin{aligned} \phi(t) &= 4\int_0^{\tan(\frac{\pi - t}{4})} [y^2 + \frac{2}{3}y^2(1-y^2)]\frac{dy}{1+y^2} \\ &= \frac{28}{3}\tan\left(\frac{\pi - t}{4}\right) - \frac{8}{9}\tan^3\left(\frac{\pi - t}{4}\right) - \frac{7}{3}(\pi - t), \end{aligned} \tag{2.43}$$

for $0 \le t \le \pi$.

Example 2.4. Consider 4 points on $\mathcal{C}$ at $s_1 = 0$, $s_2 = 0.10$, $s_3 = 0.40$ and $s_4 = 0.5$. For these points we have $t_1 = 0.1$, $t_2 = 0.3$, $t_4 = 0.1$. Thus, according to (2.33), the probability that neither one of these points is covered by a shadow arc is

$$\begin{aligned} p(0.1, 0.3, 0.1) &= 1 - \frac{4}{2\pi} \cdot 1.1141 + \frac{1}{2\pi}[2\phi(0.1) + \phi(0.3)] \\ &= 1 - 0.70926 + \frac{1}{2\pi}(2 \times 1.01588 + 0.83253) \\ &= 0.7466. \end{aligned}$$

Finally, in the Poisson random field, if $\mu = 1$, the expected number of disks (or shadow arcs) is $\lambda = \mu\pi = 3.1416$, and the probability that the above 4 points remain uncovered is

$$\begin{aligned}\psi &= E\{p(0.1, 0.3, 0.1)^N\} \\ &= \exp\{-\lambda(1 - p(0.1, 0.3, 0.1))\} \\ &= \exp\{-3.1416(1 - 0.7466)\} = 0.4511.\end{aligned}$$

■

2.8. Distribution of Clump Size In a Poisson Field On The Line

Consider a coverage process on the line, driven by a Poisson process. In this process, N (black) intervals of random length, W, are randomly placed on the line. Let $\tau_1 \leq \tau_2 \leq \cdots \leq \tau_N$ be left hand limits of the (black) intervals. We assume that τ_i $(i = 1, \cdots, N)$ are driven by a Poisson process with mean of λ per unit length. In other words, let (ξ_1, ξ_2) be any interval on the line. Let $N(\xi_1, \xi_2)$ be the number of τ values in this interval, i.e., $N(\xi_1, \xi_2) = \sum_{i=1}^{N} I\{\tau_i \in (\xi_1 \xi_2)\}$. $N(\xi_1, \xi_2)$ is a random variable having a Poisson distribution with mean $\lambda(\xi_2 - \xi_1)$. Moreover, if $\xi_1 < \xi_2 < \xi_3 < \xi_4$, $N(\xi_1, \xi_2)$ is independent of $N(\xi_3, \xi_4)$. In particular, if the N (black) intervals fall within the interval (ξ^*, ξ^{**}) then $N \equiv N(\xi^*, \xi^{**})$ has a Poisson distribution with mean $\mu^* = \lambda(\xi^{**} - \xi^*)$. Given that $\{N = n\}$, the conditional distribution of $\tau_1 < \tau_2 < \cdots < \tau_n$ is like that of the order statistics of n i.i.d. random variables, having a uniform distribution on (ξ^*, ξ^{**}). For the proof of this result, as well as for additional properties of the Poisson point process, see Karlin and Taylor (1975). Let $X_i = \tau_i - \tau_{i-1}$. The random variables $X_1, X_2, \cdots$ (interarrival timer) are **independent**, having the exponential distribution $\Pr\{X \leq x\} = 1 - \exp\{-\lambda x\}$, $0 \leq x < \infty$.

We further assume that each (black) interval has random length W, independent of its location τ, and $W_1, W_2, \cdots$ are independent and identically distributed (i.i.d.) andom variables, having a common c.d.f. $G(y)$, $a \leq y \leq b$.

If the length of the i-th (black) interval, W_i is greater than X_{i+1} $(i = 1, 2, \cdots)$ the -th and $(i+1)$st (black) intervals overlap. A **clump** is an interval on the line, which s covered by (black) intervals. Every clump is followed by a vacant interval. The ength of such a vacant interval has an exponential distribution with mean $\mu_\nu = \frac{1}{\lambda}$. An nteresting and not an easy problem is to find the distribution of the length of clumps. he reader is referred to the book of Hall (1988) pp. 89 for the distribution of the clump ength, when Y has a fixed value. A general explicit form of this distribution is not vailable. Hall (p. 89) provides the Laplace Transform of this distribution. We present ere an algorithm for the determination of the c.d.f. of the clump length, which is based n a theory of Chernoff and Daly (1957), concerning the distribution of the remaining ump length to the right of a covered point. We apply this algorithm again, in a more mplicated form, in Chapter 6, where we develop the distribution of the length of a adow on the line.

Without loss of generality, let $\xi^* = 0$. Assume that W has a continuous c.d.f., $G(w)$, th p.d.f. $g(w)$ and **finite** mean, $E\{W\} < \infty$. Suppose that a point x in the interval

$(0, \xi^{**})$ is covered by (black) interval(s). Let $N(x, y)$, for $x < y < \xi^{**}$, denote the number of (black) intervals covering **both** x and y.

Following Chernoff and Daly (1957) we define

$$T^{(0)}(x) = x \tag{2.44}$$

$$T^{(1)}(x) = \sup\{y : x \leq y, \ \text{ such that } \ N(x, y) \geq 1\}, \tag{2.45}$$

and

$$T^{(i+1)}(x) = T^{(1)}(T^{(i)}(x)), \quad i \geq 1. \tag{2.46}$$

Notice that $T^{(i+1)}(x) \geq T^{(i)}(x)$ for each $i \geq 0$ and each x, with probability one. Hence, $T^{(\infty)}(x) = \lim_{i \to \infty} T^{(i)}(x)$ is the right hand limit of a clump, to the right of the point x. We show now how the distribution of $T^{(\infty)}(x)$ can be determined.

The probability that a (black) interval covers both x and y $(x < y)$ is

$$p(x, y) = \frac{1}{\xi^{**}} \int_0^x (1 - G(y - z))dz \tag{2.47}$$

since the left hand limit, τ, has a uniform distribution on $(0, \xi^{**})$. Thus, $N(x, y)$ has a Poisson distribution with mean

$$\begin{aligned} \eta(x, y) &= \mu^{**} p(x, y) \\ &= \lambda \int_0^x (1 - G(y - z))dz, \end{aligned} \tag{2.48}$$

where $\mu^{**} = \lambda \xi^{**}$ is the expected number of (black) intervals in $(0, \xi^{**})$.

Let $H_1(t \mid x)$ be the c.d.f. of $T^{(1)}(x)$. We have the relationship, for $t \geq x$,

$$\begin{aligned} 1 - H_1(t \mid x) &= \Pr\{T^{(1)}(x) > t\} \\ &= \Pr\{N(x, t) \geq 1\} \\ &= 1 - \exp\{-\lambda \int_0^x (1 - G(t - z))dz\}. \end{aligned}$$

Thus,

$$\begin{aligned} H_1(t \mid x) &= \exp\{-\lambda \int_0^x (1 - G(t - z))dz\} \\ &= \exp\{-\lambda \int_{t-x}^t (1 - G(w))dw\}. \end{aligned} \tag{2.49}$$

It is interesting to notice that the c.d.f. $H_1(t \mid x)$, which is continuous for $t > x$, has jump point at $t = x$, given by

$$H_1(x \mid x) = \exp\{-\lambda \int_0^x (1 - G(w))dw\}. \tag{2.50}$$

$H_1(x \mid x)$ is the probability that the point x is the right hand limit of a clump. addition, since $E\{W\} = \int_0^\infty (1 - G(w))dw < \infty$, $\lim_{t \to \infty} \int_{t-x}^t (1 - G(y))dy = 0$. It follo from (2.49) that $\lim_{t\to\infty} H_1(t \mid x) = 1$. The p.d.f. of $T^{(1)}(x)$, to the right of x, is

$$h_1(t \mid x) = \lambda(G(t) - G(t-x))H_1(t \mid x), \tag{2.51}$$

for $t > x$.

The c.d.f. of $T^{(2)}(x) = T^{(1)}(T^{(1)}(x))$ is

$$\begin{aligned} H_2(t \mid x) &= H_1(x \mid x) + \int_x^t h_1(y \mid x)H_1(t \mid y)dy \\ &= H_1(x \mid x) + \lambda \int_x^t [G(y) - G(y-x)] \exp\left\{ -\lambda\left(\int_{y-x}^{y} (1 - G(w))dw \right.\right. \\ &\quad \left.\left. + \int_{t-y}^{t} (1 - G(w))dw \right) \right\} dy. \end{aligned} \tag{2.52}$$

The function $H_2(t \mid x)$ has also a jump at $t = x$ which is $H_2(x \mid x) = H_1(x \mid x)$. Generally, for every $i \geq 2$, if $H_i(t \mid x)$, $t \geq x$, is the c.d.f. of $T^{(i)}(x)$ then we obtain, recursively,

$$H_{i+1}(t \mid x) = H_1(x \mid x) + \int_x^t h_i(y \mid x)H_1(t \mid y)dy, \quad i \geq 1 \tag{2.53}$$

where $h_i(t \mid x)$ is the p.d.f. of $H_i(t \mid x)$ for $t > x$.

Since $T^{(i+1)}(x) \geq T^{(i)}(x)$, $H_{i+1}(t \mid x) \leq H_i(t \mid x)$ for each $t \geq x$ and all $i \geq 1$. Hence the c.d.f. of $T^{(\infty)}(x)$ is

$$H^*(t \mid x) = \lim_{i\to\infty} H_i(t \mid x). \tag{2.54}$$

The c.d.f. $H^*(t \mid x)$, of the right hand limit of a clump covering x, can be determined numerically, as will be shown in the following example.

Finally, let $U(x)$ denote the right hand limit of a clump starting at x. The c.d.f. of $U(x)$ is given by

$$K(u \mid x) = \int_x^u g(y-x)H^*(u \mid y)dy. \tag{2.55}$$

Notice that the length of the clump is $U(x) - x$.

Example 2.5. In the present example we illustrate the algorithm for determining the distribution of the size of a clump. We consider the case where W has an exponential distribution with $E\{W\} = 1$. In this case

$$H_1(t \mid x) = \exp\{-\lambda e^{-(t-x)}(1 - e^{-x})\}, \quad t \geq x \tag{2.56}$$

and

$$H_1(x \mid x) = \exp\{-\lambda(1 - e^{-x})\}.$$

It is straightforward to show that

$$H_2(t \mid x) = H_1(x \mid x) + \lambda(1 - e^{-x}) \int_x^t e^{-(y-x)} \exp\{-\lambda e^{-(y-x)} \cdot \\ \cdot (1 - e^{-x}) - \lambda e^{-(t-y)}(1 - e^{-y})\}dy, \quad t \geq x.$$

Formulae for $H_i(t \mid x)$ for values of i greater than 2 can be derived, but the process becomes very cumbersome. The following discrete approximation yields very fast numerical results: Fix a small value of Δ and let $t_j = x + j\Delta$ $(j = 1, 2, 3, \cdots)$. We compute the c.d.f. $H_i(t \mid x)$ at the values of t_j, recursively, according to the following approximation.

$$H_{i+1}(t_j \mid x) = H_1(x \mid x) + \sum_{l=1}^{j} [H_i(t_l \mid x) \\ - H_i(t_{l-1} \mid x)] H_1\left(t_j \mid t_l - \frac{\Delta}{2}\right); \quad j = 1, 2, \cdots, i \geq 1.$$

If λ is small the number of covering (black) intervals over a specified interval is small, and the clump size tend to be small. This is reflected in the following table, in which $\lambda = 0.1$, $x = 1$.

We see in Table 2.1 that after $k = 10$ interactions the differences between $H_9(t_j \mid 1)$ and $H_{10}(t_j \mid 1)$ are quite small. It seems that in this case $(\lambda = 0.1)$, it is sufficient to make 10 iterations to approximate $H^*(t_j \mid x)$. Finally, suppose that a clump starts at x^0 then the c.d.f. $K(u \mid x^0)$ can be approximated, in our case, by

$$\hat{K}(u_j \mid x_0) = \sum_{l=1}^{j} [\exp(-(l-1)\Delta) - \exp(-l\Delta)] \cdot H_{10}\left(u_j \mid u_l - \frac{\Delta}{2}\right),$$

where $u_i = x_0 + i\Delta$ $(i = 1, 2, \cdots)$. The values of $\hat{K}(u_j \mid x_0)$ for $x_0 = 0$ and $\Delta = .5$ are given in Table 2.2.

Table 2.1. Values of $H_i(t_j \mid 1)$ for $\Delta = 1/2$ and $\lambda = 0.1$

$j \backslash i$	1	2	8	9	10
0	0.9387	0.9387	0.9387	0.9387	0.9387
1	0.9624	0.9611	0.9548	0.9539	0.9531
2	0.9770	0.9753	0.9665	0.9652	0.9640
3	0.9860	0.9844	0.9753	0.9740	0.9726
4	0.9915	0.9901	0.9819	0.9806	0.9794
5	0.9948	0.9937	0.9869	0.9858	0.9846
6	0.9969	0.9960	0.9906	0.9896	0.9886
7	0.9981	0.9975	0.9933	0.9925	0.9917
8	0.9988	0.9984	0.9952	0.9946	0.9940
9	0.9993	0.9990	0.9967	0.9962	0.9957
10	0.9996	0.9994	0.9977	0.9973	0.9969

Table 2.1. Values of $H_i(t_j \mid 1)$ for $\Delta = 1/2$ and $\lambda = 0.1$ (Continued)

$j \backslash i$	1	2	8	9	10
11	0.9997	0.9996	0.9984	0.9981	0.9978
12	0.9998	0.9998	0.9989	0.9987	0.9984
13	0.9999	0.9998	0.9992	0.9991	0.9989
14	0.9999	0.9999	0.9995	0.9994	0.9992
15	1.0000	0.9999	0.9996	0.9996	0.9995
16	1.0000	1.0000	0.9998	0.9997	0.9996
17	1.0000	1.0000	0.9998	0.9998	0.9997
18	1.0000	1.0000	0.9999	0.9999	0.9998
19	1.0000	1.0000	0.9999	0.9999	0.9999
20	1.0000	1.0000	0.9999	0.9999	0.9999
21	1.0000	1.0000	1.0000	1.0000	0.9999
22	1.0000	1.0000	1.0000	1.0000	1.0000
23	1.0000	1.0000	1.0000	1.0000	1.0000

Table 2.2. Values of $\hat{K}(u_j \mid 0)$, $\Delta = .5$, $\lambda = 0.1, 0.5$

	$\hat{K}(U_i$	$\mid 0)$		$\hat{K}(U_i$	$\mid 0)$
i	$\lambda = 0.1$	$\lambda = 0.5$	i	$\lambda = 0.1$	$\lambda = 0.5$
1	0.3849	0.3523	11	0.9896	0.8761
2	0.6144	0.5389	12	0.9927	0.8914
3	0.7544	0.6450	13	0.9949	0.9057
4	0.8412	0.7094	14	0.9964	0.9189
5	0.8958	0.7515	15	0.9975	0.9310
6	0.9307	0.7815	16	0.9983	0.9419
7	0.9534	0.8049	17	0.9988	0.9515
8	0.9683	0.8249	18	0.9991	0.9599
9	0.9782	0.8431	19	0.9964	0.9671
10	0.9850	0.8600	20	0.9996	0.9733

3

Visibility Probabilities

In the previous chapter we discussed various coverage problems. In the present chapter we start the development of the theory of **stochastic visibility in random fields**.

Problems of stochastic visibility are typically problems whether a line of sight from the origin **O** reaches a target **T**, without being interfered (intersected) by randomly placed objects in the field between **O** to **T**. Problems of stochastic visibility can be translated to coverage problems by presenting the line of sight as a light ray from **O** toward the target **T**. If the light ray is not intersected by an obscuring object in the field then **T** is in the light (visible), otherwise **T** is in the shadow (invisible), which is cast by the obscuring objects. The random obscuring objects in the field cast shadows of random size on the targets, which are either points on lines or line segments in the plane or in three dimensional spaces. We model the obscuring elements as random disks in the plane, or random spheres in three dimensional spaces.

In the present chapter we develop geometric and analytical methods for determining the probabilities that specified points, or intervals, in the plane are visible from a single observation point. In Chàpter 4 we extend the methods to cases of several observation points. The fields of obscuring elements are either standard Poisson fields, non-standard Poisson fields or multinomial fields.

3.1. Geometric Methods: Standard Poisson Fields

3.1.1. The Targets and Observation Points Are Within the Scattering Region

Let **O** and **T** designate, respectively the observation and the target points. Both **O** and **T** are points *within* a region, $\mathcal{S}$, in which random obscuring disks are scattered. We assume first that both **O** and **T** can be covered by an obscuring disk and develop the formula of the visibility probability, ψ, of **T** from **O** under a standard Poisson field of intensity λ. For a Poisson field, we have to compute the expected number of obscuring disks, $\eta(T) = E\{N\{T\}\}$ which can intersect the line segment $\overline{OT}$. The visibility probability is $\psi = \exp\{-\eta(T)\}$.

Let d be the distance between **O**, and **T**. Let $B(Y)$ denote the set of all centers of disks of radius Y, which intersect $\overline{OT}$. (Figure 3.1). The area of this set is

$$A\{B(Y)\} = \iint_{B(Y)} dx_1 dx_2 = 2dY + \pi Y^2. \qquad (3.1)$$

Let μ_1 and μ_2 denote the first and second moments of the distribution of Y. Since the field is standard, the distribution of Y does not depend on the centers (x_1, x_2) of the disks, and the expected number of disks which can intersect $\overline{OT}$ is

$$\eta(T) = \lambda(2d\mu_1 + \pi\mu_2). \qquad (3.2)$$

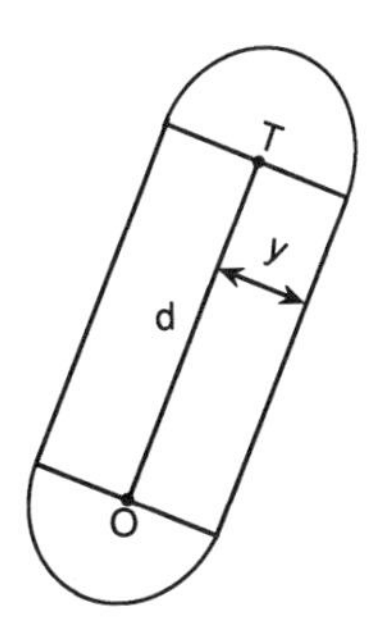

Figure 3.1. The Set $B(Y)$ of Centers of Disks of Radius, Y, Which Intersect $\overline{OT}$.

Example 3.1. Suppose that both **O** and **T** are within a scattering region. The random disks from a standard Poisson field with $\lambda = 0.1[1/\text{m}^2]$. The radii of random disks have a uniform distribution over the interval $[0.3, 0.6][\text{m}]$ and the distance between **O** and **T** is $d = 200[\text{m}]$. Both **O** and **T** can be covered by random disks. It is easy to check that

$$\mu_1 = \frac{1}{2}(0.3 + 0.6) = 0.45[\text{m}]$$

and

$$\mu_2 = \frac{1}{12}(0.6 - 0.3)^2 + \frac{1}{4}(0.3 + 0.6)^2 = 0.21[\text{m}^2].$$

According to (3.2), the expected number of disks that can intersect $\overline{OT}$ is

$$\eta(T) = 0.1[2 \cdot 200 \cdot 0.45 + \pi \cdot 0.21] = 18.066.$$

The visibility probability of **T** from **O** is $\psi = e^{-18.066} \approx 0$. If, on the other hand, $\lambda = .01[1/\text{m}^2]$ then $\eta(T) = 1.8066$ and the visibility probability is $\psi = e^{-1.8066} = .1642$. ■

Example 3.2. The scattering region consists of two parallel strips, $\mathcal{S}_1$ and $\mathcal{S}_2$ (see Figure 3.2). **O** is within $\mathcal{S}_1$, having coordinates $(0,0)$. **T** is within $\mathcal{S}_2$, with coordinates (x_t, y_t). The distance between $\mathcal{S}_1$ and $\mathcal{S}_2$ is D. The Poisson field over $\mathcal{S}_1$ is standard, with intensity λ_1. The Poisson field over $\mathcal{S}_2$ is standard with intensity λ_2.

We assume that the distribution of radii is the same over $\mathcal{S}_1$ and over $\mathcal{S}_2$. The distance of **O** from the boundary $\mathcal{S}_1$ is l_1 and l_2 is the distance of **T** from the boundary $\mathcal{S}_2$. Let β be the slope of $\overline{OT}$ and $\theta = \tan^{-1}(\beta)$. Verify that $h = 2Y/\beta$. Let $B_1(Y)$ be the set in $\mathcal{S}_1$ of disk centers of radius Y, which intersect $\overline{OT}$. Let $B_2(Y)$ be the corresponding set in $\mathcal{S}_2$. Elementary geometry yields that the area of $B_i(Y)$, $i = 1, 2$, is

$$A_i(Y) = 2\frac{l_i}{\sin\,\theta}Y + \frac{\pi}{2}Y^2. \tag{3.3}$$

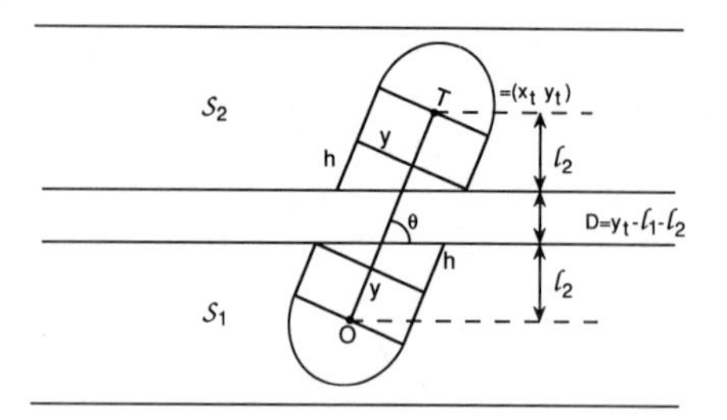

Figure 3.2. The Geometry of $B(Y)$.

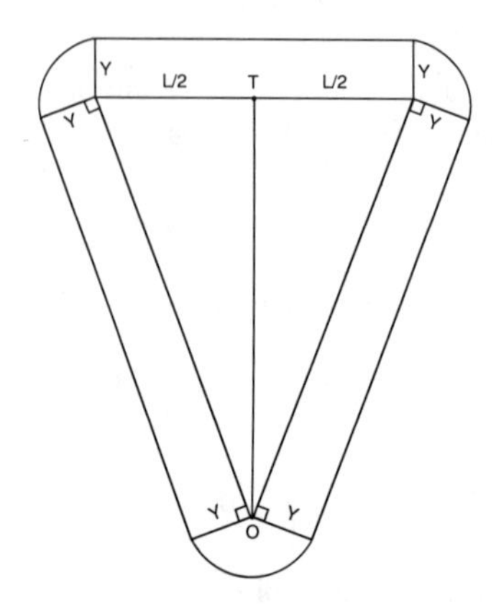

Figure 3.3. The set $B_L(Y)$.

It follows that the visibility probability of **T** from **O** is

$$\psi = \exp\left\{-\lambda_1\left(2\frac{l_1}{\sin\theta}\mu_1 + \frac{\pi}{2}\mu_2\right) - \lambda_2\left(2\frac{l_2}{\sin\theta}\mu_1 + \frac{\pi}{2}\mu_2\right)\right\}. \tag{3.4}$$

Notice that formula (3.4) is based on the assumption that the visibility within S_1 is independent of that within S_2.

For identifying a target, it is generally insufficient that a single point **T** is visible. It is often required that a whole line segment of length L centered at **T** will be observable. In Figure 3.3 we present the set $B_L(Y)$, of possible centers of disks of radius Y which obscure parts or whole of the line interval $(\mathbf{T} - \frac{L}{2}, \mathbf{T} + \frac{L}{2})$. It is straight forward to generalize Eq. (3.2) and show that the expected number of disks which may obscure the whole or part of the interval around **T**, when d is the distance of **T** from **O**, is

$$\eta_L(\mathbf{T}) = \lambda\frac{Ld}{2} + \lambda\left(L + 2d\left(1 + \left(\frac{L}{2d}\right)^2\right)^{1/2}\right)\mu_1 + \lambda\pi\mu_2. \tag{3.5}$$

The probability that a line segment (window) of size w around **T** will be completely observable is

$$\psi(\mathbf{T}; W) = \exp\{-\eta_L(T)\}. \tag{3.6}$$

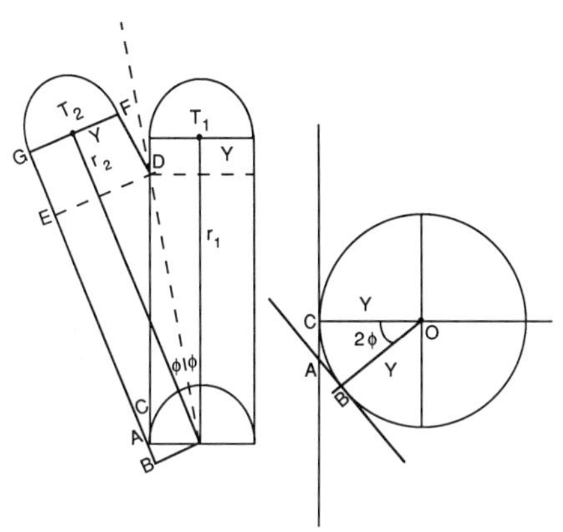

Figure 3.4. The Geometry of $B(Y)$ in Case 1

We derive now the conditional visibility probability of several target points, given that both the observation point is not covered and the target points are within the scattering region. We introduce a polar coordinate system with the origin at **O**. Thus, every target point is specified by its polar coordinates (r, s), where $0 \le r < \infty$ is the distance of **T** from **O** and s, $-\pi < s \le \pi$, is its orientation angle. Without loss of generality, we assign the vertical axis (the North) an orientation $s = 0$ and the horizontal axis (the East) an orientation $s = \pi/2$. We order the k target points $\mathbf{T}_i = (r_i, s_i)$, $i = 1, \cdots, k$, in descending order of their orientation coordinates, (counter clockwise), i.e.,

$$\pi \ge s_1 > s_2 > \cdots > s_k > -\pi.$$

We develop first the formula of the simultaneous visibility probability of two target points, $k = 2$, and later generalize it for any k points.

Let $\phi = \frac{1}{2}(s_1 - s_2)$ and let $X = Y \cot(\phi)$, where Y is the random radius of a disk. The geometry of the set $B(Y)$, of all centers of disks of radius Y, which intersect either $\overline{OT_1}$ or $\overline{OT_2}$, depends on three cases.

Case 1: $X \le \min(r_1, r_2)$;

Case 2: $\min(r_1, r_2) < X \le \max(r_1, r_2)$;

Case 3: $X > \max(r_1, r_2)$.

In practical situations Case 1 is prevalent. For example, if $min(r_1, r_2) = 50$[m] and the distribution of Y is concentrated on $[a, b]$, where $b = 1$[m] then $X \le 1 \cot(\phi)$[m]. Thus, Case 1 prevails if $1 \cdot \cot\phi \le 50$ or $\phi \ge \tan^{-1}(1/50) = 0.02[1.146°]$.

Although one has to develop the theory for all $0 < \phi \le \pi/2$, we will restrict attention in the present section to Case 1, in which $\phi \ge \tan^{-1}(b/\min(r_1, r_2))$. A general formula will be derived in Section 3.3.

The simultaneous visibility probability of $\mathbf{T}_1$ and $\mathbf{T}_2$ from **O** is

$$\psi_2(\mathbf{r}^{(2)}, \phi) = \exp[-\lambda E\{A_2(Y; \mathbf{r}^{(2)}, \phi)\}], \tag{3.7}$$

where $A_2(Y; \mathbf{r}^{(2)}, \phi)$ is the area of $B(Y)$; $\mathbf{r}^{(2)} = (r_1, r_2)$.

We write the area of $B(Y)$ as

$$A_2(Y; \mathbf{r}^{(2)}, \phi) = A_1(Y; r_1) + \Delta_2(Y; r_2, \phi), \tag{3.8}$$

in which $A_1(Y;r_1) = 2Yr_1$ is the area of the set of centers of disks of radius Y which can intersect $\overline{OT}_1$. Notice that due to the assumption that $\mathbf{O}$ is not covered, we have to subtract the area of a half circle of radius Y above $\mathbf{O}$. This compensates for the area of the half circle above $\mathbf{T}_1$. The term $\Delta_2(Y;r_2,\phi)$ represents the area of the set of centers of disks which can intersect $\overline{OT}_2$, and is disjoint of the set of centers of disks which can intersect $\overline{OT}_1$. This area, $\Delta_2(Y;r_2,\phi)$, is the sum of several areas. One is the area of the half circle above $\mathbf{T}_2$, $(\pi/2)Y^2$. Second area is that of the rectangle $EDFG$, which is given by $2Y(r_2 - Y\cot\phi)$. Third area is that of the triangle ADE. Notice that $\overline{AE} = 2Y\cot(2\phi)$. Hence, the area of this triangle is $2Y^2\cot(2\phi)$. The fourth area is that of the rhombus $OCAB$, minus the area of the sector of the circle of radius Y centered at $\mathbf{O}$ with angle 2ϕ, i.e., ϕY^2. Notice that the rhombus $OCAB$ consists of two congruent right triangles OCA and OBA. Moreover, $\overline{CA} = \overline{AB} = Y\tan\phi$. Hence, the area of the rhombus $OCAB$ is $Y^2\tan\phi$. Adding all these components we obtain

$$\Delta_2(Y;r_2,\phi) = 2r_2Y + \left[\left(\frac{\pi}{2} - \phi\right) - \frac{1}{\tan(\phi)}\right]Y^2 \tag{3.9}$$

Finally, in Case 1, i.e., for $\phi > \tan^{-1}(b/\min(r_1,r_2))$,

$$E\{A_2(Y;\mathbf{r}^{(2)},\phi)\} = 2(r_1+r_2)\mu_1 + \left(\frac{\pi}{2} - \phi - \frac{1}{\tan(\phi)}\right)\mu_2 \tag{3.10}$$

Notice that $E\{A_2(Y;\mathbf{r}^{(2)},\pi/2)\} = 2(r_1+r_2)\mu_1$.

Example 3.3. Two targets are at distance of 100[m] from $\mathbf{O}$ and the angle between $\overline{OT}_1$ and $\overline{OT}_2$ is $s_1 - s_2 = 15°$. Thus $\phi = .131$ [radians]. The intensity of the field is $\lambda = .01$ [1/m^2]. The radii of the disks are uniformly distributed over (.3, .6)[m]. Thus, $\mu_1 = .45$[m] and $\mu_2 = .21$[m^2]. Since $\phi > \tan^{-1}(.6/100) = .006$ we can apply formula (3.10) of Case 1. Thus we obtain $E\{A_2(Y;\mathbf{r},\phi)\} = 178.71$[m^2], and the simultaneous visibility probability is $\psi_2(\mathbf{r}^{(2)},\phi) = \exp(-1.787) = .1674$. If the two targets are only 50[m] from $\mathbf{O}$ then the visibility probability increases to $\psi_2(\mathbf{r}^{(2)},\phi) = \exp(-.887) = .4119$. ■

Formula (3.10) can now be generalized for k targets, provided we remain in Case 1 for all k targets. More precisely, let $\phi_i = \frac{1}{2}(s_i - s_{i+1})$, $i = 1,\cdots,k-1$ and assume that

$$\min_{1\le i\le k-1} \phi_i \ge \tan^{-1}\left(\frac{b}{\min\limits_{1\le i\le k}(r_i)}\right) \tag{3.11}$$

By piecewise addition of area increments we obtain the general formula

$$\begin{aligned} E\{A_k(Y;\mathbf{r}^{(k)},\phi^{(k-1)})\} &= 2\mu_1\sum_{i=1}^{k} r_i \\ &+ \mu_2\sum_{i=1}^{k-1}\left[\frac{\pi}{2} - \phi_i - 1/\tan(\phi_i)\right]. \end{aligned} \tag{3.12}$$

Here $\mathbf{r}^{(k)} = (r_1, \cdots, r_k)$, $\phi^{(k-1)} = (\phi_1, \cdots, \phi_{k-1})$. The corresponding simultaneous visibility probability is

$$\psi_k(\mathbf{r}^{(k)}, \phi^{(k-1)}) = \exp(-\lambda E\{A_k(Y; \mathbf{r}^{(k)}, \phi^{(k-1)}\}). \tag{3.13}$$

Example 3.4. We consider $k = 4$ target points located on the circumference of a half-circle of radius $r = 50[\mathrm{m}]$ centered at $\mathbf{O}$. The orientation angles of these points are: $s_1 = 75°$, $s_2 = 35°$, $s_3 = -15°$ and $s_4 = -85°$. The values of $\phi_i = (s_i - s_{i+1})/2$ (in radians) and the expected incremented areas, $E\{\Delta_{i+1}(Y; \mathbf{r}, \phi_i)\}$, with $\mu_1 = .45[\mathrm{m}]$ and $\mu_2 = .21[\mathrm{m}^2]$ are given in the following table.

i	ϕ_i	$E\{\Delta_{i+1}(Y; \mathbf{r}, \phi_i)\}$
1	.3491	44.6796
2	.4363	44.7879
3	.6109	44.9017
sum	—	134.3692

The expected total area of $B(Y)$ is according to Eq. (3.12), $134.3692 + 8\mu_1 r = 314.3692$.

Accordingly, if $\lambda = .001[1/\mathrm{m}^2]$ then $\psi_4(\mathbf{r}, \phi) = \exp\{-.31437\} = .7302$, and if $\lambda = .01[1/\mathrm{m}^2]$ then $\psi_4(\mathbf{r}, \phi) = 0.0431$. ■

Also in the multi-target problem we have to consider the joint visibility of whole intervals (windows) around the target points. We will develop a formula for the probabilities of joint visibility of intervals later in this chapter.

3.1.2. The Targets and the Observation Points Are Outside a Rectangular Scattering Region

In the present section we derive formulae for the visibility probabilities, when the targets are on one side of a rectangular scattering region and the observation point on ts other side. The scattering region under consideration is a horizontal rectangle with wo parallel sides of distance u and w, $0 < u < w$, from the observation point (the rigin) $\mathbf{O}$. These horizontal boundaries of the region are of very large (infinite) length. The polar coordinates of the target points $\mathbf{T}_i$ are (r_i, s_i), $i = 1, \cdots, k$. The radii of lisks are distributed over $[a, b]$, and $r_i > (w + b)/\cos(s_i)$, $i = 1, \cdots, k$. Thus, no target oint can be covered by a random disk. Also in the present section we assume that the Poisson field is a standard one, with intensity λ. Thus, for the computation of visibility robabilities we have to determine areas of sets $B(Y)$ of centers of disks of radius Y, which can intersect lines of sight.

Suppose that the centers of random disks are scattered in a strip, $\mathcal{S}$, bounded between wo parallel lines of distance u and w from $\mathbf{O}$ (see Figure 3.5).

The length of the portion of $\overline{OT}$ which is within $\mathcal{S}$ is $\rho(s) = (w-u)/\cos(s)$. It follows at the area of the parallelogram $B(Y)$ is $2Y(w-u)/\cos(s)$. The visibility probability , accordingly

$$\psi_1(s) = \exp\{-2\lambda(w-u)\mu_1/\cos(s)\}. \tag{3.14}$$

'e extend this result now to the case of a line segment of length L, centered at $\mathbf{T}$, d perpendicular to $\overline{\mathbf{OT}}$ should be completely visible. The line segment is partially or

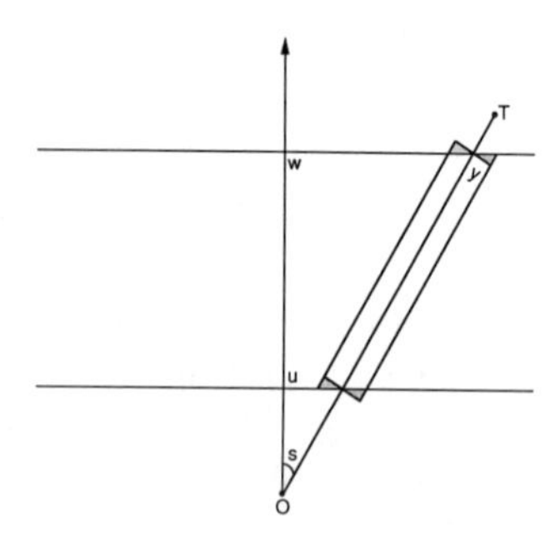

Figure 3.5. The Geometry of $B(Y)$ for One Target

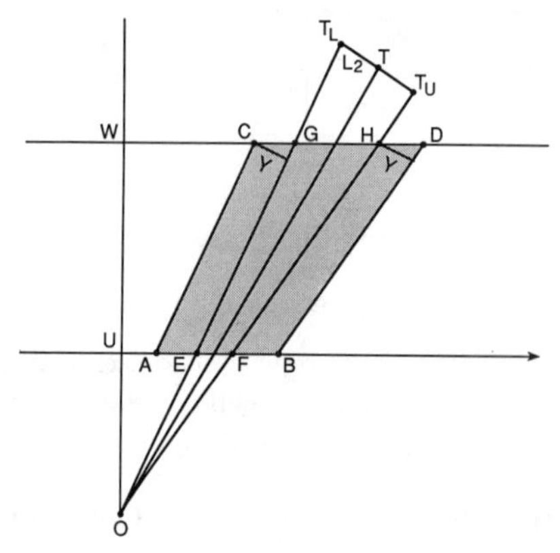

Figure 3.6. The Geometry of $B_L(Y)$

completely obscured by disks of radius Y, if their centers fall in the set $B_L(Y)$, presented in Figure 3.6.

The set $B(Y)$ is represented by the trapez $ABCD$. Let (ρ, s) be the polar coordinates of $\mathbf{T}$. Let s_L be the orientation angle of $\overline{\mathbf{OT}}_L$ and s_U the orientation angle of $\overline{\mathbf{OT}}_U$. It is immediate to verify that

$$s_L = s - \tan^{-1}\left(\frac{L}{2s}\right),$$

$$s_U = s + \tan^{-1}\left(\frac{L}{2s}\right).$$

Furthermore, the area of the trapez $EFGH$ is

$$\frac{w^2 - u^2}{2}(\tan(s_U) - \tan(s_L)).$$

The area of the parallelogram $AECG$ is $\dfrac{Y}{\cos(s_L)}(w - u)$, and the area of $FBHD$

$\dfrac{Y}{\cos(s_U)}(w-u)$. Accordingly,

$$E\{B_L(Y)\} = \frac{w^2-u^2}{2}(\tan(s_U)-\tan(s_L)) + (w-u)\left(\frac{1}{\cos(s_L)}+\frac{1}{\cos(s_U)}\right)\mu_1. \tag{3.15}$$

Finally, the probability that $\overline{\mathbf{T}_L\mathbf{T}_U}$ is completely visible is

$$\psi_L(s) = \exp\left\{-\lambda\frac{w^2-u^2}{2}(\tan(s_U)-\tan(s_L)) - \lambda(w-u)\frac{\cos(s_L)+\cos(s_U)}{\cos(s_L)\cos(s_U)}\mu_1\right\}. \tag{3.16}$$

Example 3.5. A target is detectable if at least $L = 1$[m] of its front is completely visible. A wooded area (forest) starts at $u = 50$[m] from an observation point, $\mathbf{O}$, which is located at the origin. The center of the target $\mathbf{T}$ is at the point $(\rho, 30°)$. The trees are represented by random disks in a standard Poisson field with $\lambda = 0.01[1/\mathrm{m}^2]$, and random radius Y which is uniformly distributed in (.3,.6). How far away from $\mathbf{O}$ should the target be placed within the forest, so that the probability that it will be completely visible will not exceed the value of $\psi_L(s) = 0.1$?

In the present case we have to find the value of ρ (distance from $\mathbf{O}$). Since the target is within the forest, we take for w, the distance from $\mathbf{O}$ of a horizontal line which goes through $\mathbf{T}_U$.

The distance of $\mathbf{T}_U$ from $\mathbf{O}$ is $\rho_U = (\rho^2 + L^2/4)^{1/2}$. Hence

$$w(\rho) = \rho_L \cos(s_U) = (\rho^2 + L^2/4)^{1/2} \cos(s_U).$$

For $s = \dfrac{\pi}{6}$ [radians] and $L = 1$[m], $\mu_1 = 0.45$[m], $\lambda = 0.01[\mathrm{m}^2]$ substituting in Eq. (3.16), $s_L(\rho) = \dfrac{\pi}{6} - \tan^{-1}\left(\dfrac{1}{2\rho}\right)$, $s_U(\rho) = \dfrac{\pi}{6} + \tan^{-1}\left(\dfrac{1}{2\rho}\right)$ and $w(\rho) = \left(\rho^2 + \dfrac{1}{4}\right)^{1/2} \cos(s_U(\rho))$, we obtain the following values of $\psi_L(s)$,

Table 3.1. Visibility Probability As Function of Distance [m]

ρ	$\psi_L(s)$
60	0.964
80	0.679
100	0.493
120	0.362
140	0.268
160	0.200
180	0.149
200	0.117
210	0.097

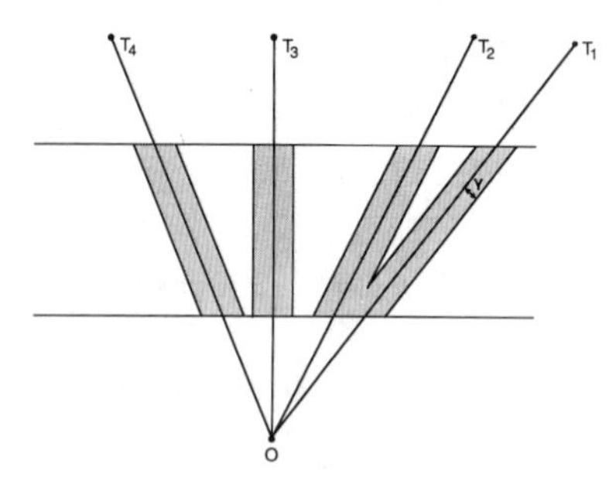

Figure 3.7. The Geometry of $B(Y)$ for Several Targets

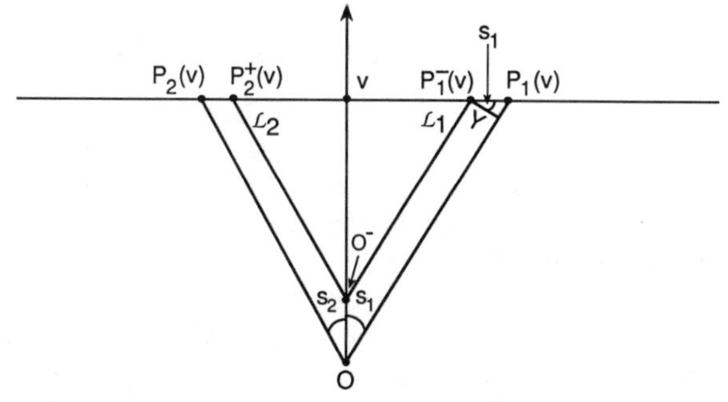

Figure 3.8. Location of the Intersection Point $\mathbf{O}^-$

We see in Table 3.1 that the required distance is about 210[m].

The set $B(Y)$ in the case of several targets may look as in Figure 3.7.

We order the target points according to their orientation points. Let $A_k(Y; s_1, \cdots, s_k)$ denote the area of $B(Y)$, corresponding to k target points, with orientation angles $s_1, \cdots, s_k$. For every $k \geq 2$ let

$$A_k(Y; s_1, \cdots, s_k) = A_{k-1}(Y; s_1, \cdots, s_{k-1}) + \Delta_k(Y; s_{k-1}, s_k). \tag{3.17}$$

$\Delta_k(Y; s_{k-1}, s_k)$ is the incremented area of the disjoint portion of $B(Y)$, due to the k-th target. We develop below the formula for the determination of $E\{\Delta_k(Y; s_{k-1}, s_k)\}$ for any $k \geq 2$. Obviously, the visibility probability of all the k targets simultaneously $\psi_k(s_1, \cdots, s_k)$ is given by

$$\psi_k(s_1, \cdots, s_k) = \psi_{k-1}(s_1, \cdots, s_{k-1}) \cdot \exp\{-\lambda E\{\Delta_k(Y; s_{k-1}, s_k)\}\}, \tag{3.18}$$

$k \geq 2$, where $\psi_1(s_1)$ is given by (3.14). In this manner one can compute the visibility probabilities $\psi_k(s_1, \cdots, s_k)$ recursively. We develop the formula of $E\{\Delta_k(Y; s_{k-1}, s_k)\}$ for $k = 2$. The general formula is obtained by replacing s_1 by s_{k-1} and s_2 by s_k.

Let $\mathcal{L}_1$ be a parallel line to $\overline{OT}_1$, on its left hand side, at distance Y from it. Let $\mathcal{L}_2$ be parallel to $\overline{OT}_2$ on its right, at distance Y. These two lines intersect each other above $\mathbf{O}$, at a point $\mathbf{O}^-$, having rectangular coordinates (x^-, u^-) (see Figure 3.8).

Let v be any level, $u < v < w$, and $\mathbf{P}_1(v)$, $\mathbf{P}_2(v)$ the points at the intersections of $\overline{OT}_1$ and $\overline{OT}_2$ with the horizontal line at a level v. The rectangular coordinates of $\mathbf{P}_i($

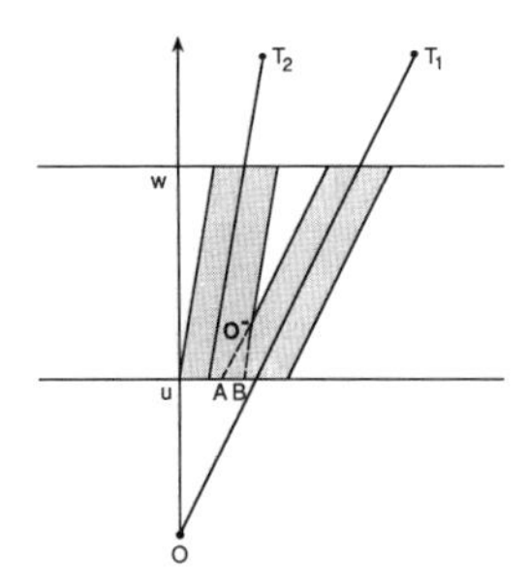

Figure 3.9. The Geometry of $B(Y)$ in Case II

are $(v \tan(s_i), v)$, $i = 1, 2$. Let $\mathbf{P}_1^-(v)$ and $\mathbf{P}_2^+(v)$ be the corresponding points on $\mathcal{L}_1$ and $\mathcal{L}_2$, respectively. The distance of $\mathbf{P}_1^-(v)$ from $\mathbf{P}_1(v)$ is $Y/\cos(s_1)$, and that of $\mathbf{P}_2^+(v)$ from $\mathbf{P}_2(v)$ is $y/\cos(s_2)$. Accordingly, the lines $\mathcal{L}_1$ and $\mathcal{L}_2$ have slopes $b_i = \cot(s_i)$, $i = 1, 2$; and intercepts $a_i = (-1)^{i-1} Y/\sin(s_i)$, $i = 1, 2$. It follows that the rectangular coordinates of the intersection point $\mathbf{O}^-$ are

$$x^- = Y \frac{\sin(s_1) + \sin(s_2)}{D(s_1, s_2)}, \tag{3.19}$$

and

$$u^- = Y \frac{\cos(s_1) + \cos(s_2)}{D(s_1, s_2)} \tag{3.20}$$

where

$$D(s_1, s_2) = \sin(s_1)\cos(s_2) - \cos(s_1)\sin(s_2). \tag{3.21}$$

We distinguish between three cases:

Case I: $u^- \leq u$;

Case II: $u < u^- \leq w$;

Case III: $w < u^-$.

Let $\alpha = (\cos(s_1) + \cos(s_2))/D(s_1, s_2)$, and let $\xi = u/\alpha$, $\eta = w/\alpha$. Cases I, II, or III hold according to whether $Y \leq \xi$, $\xi < Y \leq \eta$, or $Y > \eta$, respectively. We develop now the formula of $\Delta_2(Y; s_1, s_2)$, for each one of these three cases.

Case I: $Y \leq \xi$

When $Y \leq \xi$, the set of centers of disks or radius Y, which can intersect $\overline{OT_2}$ is similar to the set which can intersect $\overline{OT_1}$ (see Figure 3.6.) Hence, in Case I,

$$\Delta_2(Y; s_1, s_2) = 2Y(w - u)/\cos(s_2). \tag{3.22}$$

Case II: $\xi < Y \leq \eta$

In Case II,

$$\Delta_2(Y; s_1, s_2) = 2Y(w - u)/\cos(s_2) - AT(Y), \tag{3.23}$$

where $AT(Y)$ is the area of the triangle O^-AB (see Figure 3.9).

The length of the $\overline{AB}$ is $Y\gamma_2 - u\beta_2$ where

$$\beta_2 = \tan(s_1) - \tan(s_2), \tag{3.24}$$

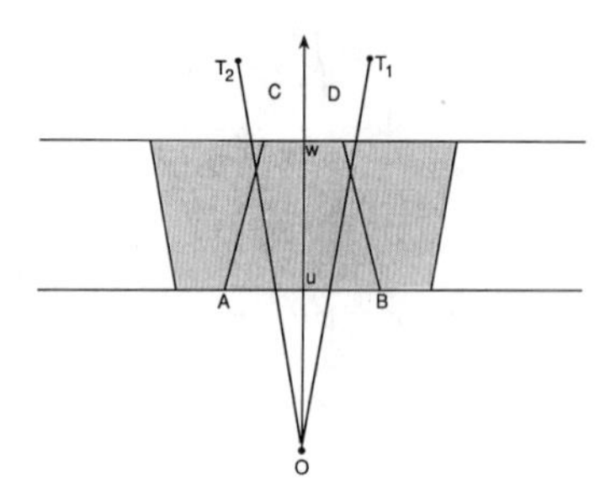

Figure 3.10. The Geometry of $B(Y)$ in Case III

and

$$\gamma_2 = \frac{1}{\cos(s_1)} + \frac{1}{\cos(s_2)}. \tag{3.25}$$

Notice that $\alpha_2\beta_2 = \gamma_2$. Since the height of the triangle O^-AB is $u^- - u$, we obtain that its area is

$$AT(Y) = \frac{1}{2}(\alpha_2\gamma_2 Y^2 - 2u\gamma_2 Y + u^2\beta_2). \tag{3.26}$$

Thus,

$$\Delta_2(Y; s_1, s_2) = \frac{u^2\beta_2}{2} + 2\left(\frac{w-u}{\cos(s_2)} + \frac{1}{2}u\gamma_2\right)Y - \frac{\alpha_2\gamma_2}{2}Y^2. \tag{3.27}$$

Case III: $Y \geq \eta$

In the third case,

$$\Delta_2(Y; s_1, s_2) = 2Y\frac{w-u}{\cos(s_2)} - ATR(Y), \tag{3.28}$$

where $ATR(Y)$ is the area of the trapezoid $ABCD$ (see Figure 3.10). It is straightforward to verify that

$$\Delta_2(Y; s_1, s_2) = \frac{w_2 - u_2}{2}\beta_2 - \delta_2(w-u)Y, \tag{3.29}$$

where

$$\delta_2 = \frac{1}{\cos(s_1)} - \frac{1}{\cos(s_2)}. \tag{3.30}$$

Changing $\alpha_2, \beta_2, \gamma_2, \delta_2, \xi, \eta$ to $\alpha_k, \beta_k, \gamma_k, \delta_k, \xi_k, \eta_k$ and replacing s_1 and s_2 with s_{k-1} an s_k, respectively, we obtain from (3.22), (3.27) and (3.29),

$$\begin{aligned} E\{\Delta_k(Y; s_{k-1}, s_k)\} &= 2\frac{w-u}{\cos(s_k)}\int_0^{\xi_k} y dF(y) - \frac{u^2\beta_k}{2}\int_{\xi_k}^{\eta_k} dF(y) \\ &+ 2\left(\frac{w-u}{\cos(s_k)} + \frac{u}{2}\gamma_k\right)\int_{\xi_k}^{\eta_k} y dF(y) - \frac{\alpha_k\gamma_k}{2}\int_{\xi_k}^{\eta_k} y^2 dF(y) \\ &+ \frac{w^2-u^2}{2}\beta_k(1-F(\eta_k)) - \delta_k(w-u)\int_{\eta_k}^{\infty} y dF(y), \end{aligned} \tag{3.31}$$

where $F(y)$ is the distribution function of Y. Since $F(y)$ is concentrated on the interval $[a, b]$, the formula of $E\{\Delta_k(Y; s_{k-1}, s_k)\}$ assumes different expressions, according to the positions of ξ_k and η_k relative to a and b. These expressions are:

(i) When $\xi_k < \eta_k \leq a$,

$$E\{\Delta_k(Y; s_{k-1}, s_k)\} = \frac{w^2-u^2}{2}\beta_k - \delta_k(w-u)\mu_1, \tag{3.32}$$

where $\mu_1 = E\{Y\}$.

(ii) When $\xi_k < a < \eta_k \leq b$,

$$\begin{aligned} E\{\Delta_k(Y; s_{k-1}, s_k)\} &= \frac{w^2-u^2}{2}\beta_k - \frac{w^2}{2}\beta_k F(\eta_k) \\ &- \delta_k(w-u)\mu_1 + \gamma_k w\mu_1(\eta_k) - \gamma_k\frac{\alpha_k}{2}\mu_2(\eta_k), \end{aligned} \tag{3.33}$$

where

$$\mu_i(\eta_k) = \int_a^{\eta_k} y^i dF(y), \qquad i = 1, 2 \tag{3.34}$$

is the partial moment of the i-th order.

(iii) When $\xi_k < a < b < \eta_k$,

$$E\{\Delta_k(Y; s_{k-1}, s_k)\} = 2\left(\frac{w-u}{\cos(s_k)} + \frac{u}{2}\gamma_k\right)\mu_1 - \frac{u^2\beta_k}{2} - \frac{\alpha_k\gamma_k}{2}\mu_2. \tag{3.35}$$

(iv) When $a \leq \xi_k < \eta_k \leq b$

$$\begin{aligned} E\{\Delta(Y; s_{k-1}, s_k)\} &= \gamma_k(w\mu_1(\eta_k) - u\mu_1(\xi_k)) \\ &- \frac{1}{2}\alpha_k\gamma_k(\mu_2(\eta_k) - \mu_2(\xi_k)) \\ &+ \frac{w^2\beta_k}{2}(1-F(\eta_k)) - \frac{u^2\beta_k}{2}(1-F(\xi_k)) - \delta_k(w-u)\mu_1. \end{aligned} \tag{3.36}$$

(v) When $a \leq \xi_k < b < \eta_k$,

$$\begin{aligned} E\{\Delta_k(Y, s_{k-1}, s_k)\} &= 2w\mu_1/\cos(s_k) \\ &+ u\delta_k(\mu_1 - \mu_1(\xi_k)) - \frac{u^2\beta_k}{2}(1-F(\xi_k)) \\ &- \frac{1}{2}\alpha_k\gamma_k(\mu_2 - \mu_2(\xi_k)). \end{aligned} \tag{3.37}$$

Finally,

(vi) When $b \leq \xi_k$

$$E\{\Delta_k(Y; s_{k-1}, s_k\} = 2(w-u)\mu_1/\cos(s_k). \tag{3.38}$$

Example 3.6. Consider a scattering strip, $\mathcal{S}$, with $u = 50$[m], $w = 75$[m]. The observation point is at the origin and $k = 5$ target points are on the other side of the strip, at orientation angles 45°, 44°, 39°, 0°, −1°. The radii of disks have uniform distributions on (0.5, 1.0). The intensity of the Poisson field is $\lambda = .001[1/\text{m}^2]$.

In the case of uniform distribution Y, over (a, b), the partial moments are

$$\begin{aligned}\mu_1(\eta) &= \frac{1}{2}(\eta_2 - a^2)/(b-a), \quad a \leq \eta < b \\ &= (b+a)/2, \quad \eta \geq b;\end{aligned}$$

and

$$\mu_2(\eta) = \begin{cases} \frac{1}{3}(\eta^3 - a^3)/(b-a), & a \leq \eta < b \\ \\ \frac{1}{3}(a^2 + ab + b^2), & b \leq \eta. \end{cases}$$

Program, "SIMVP" can be used to compute the simultaneous visibility probabilities $\psi_k(s_1, s_2, \cdots, s_k)$, for $k \geq 1$. According to this program $EDN = \exp\{E\{\Delta_k(Y, s_{k-1}, s_k)$ and $VP = \psi_k$, $k \geq 1$ ("VP")

k	s°	EDN	VP
1	45	—	.9483
2	44	.9534	.9042
3	39	.9529	.8616
4	0	.9632	.8299
5	-1	.9734	.8078

The value of VP for $k = i$ ($i = 1, \cdots, 5$) is the simultaneous visibility probability fo the first i target points, with coordinates $s_1 > \cdots > s_i$.

3.2. Analytic Methods: General Poisson Fields

In the present section we develop a general analytical method for determining visibilit probabilities, when the observation point and the targets are outside the scatterir regions. The scattering region can assume a variety of different shapes and the Poissc field may be standard or general.

3.2.1. The General Theoretical Framework

We use in this section a system of polar coordinates. Thus, let the observation point be the origin (0,0), and let $\mathbf{T}_i = (r_i, s_i)$ $i = 1, \cdots, k$ be the k targets ($k \geq 1$).

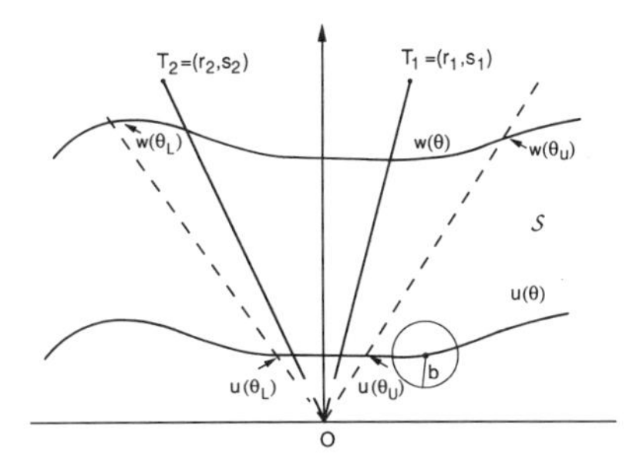

Figure 3.11. The Geometry of the Field.

A star-shaped function $v(\theta)$, $-\pi < \theta \le \pi$, represents the distance from **O**, of a point having orientation angle θ. All orientation angles are measured from the vertical axis, and the horizontal axis has orientation $\theta = \pi/2$, (see Figure 3.11).

The scattering strip, $\mathcal{S}$, is specified by two boundary functions $u(\theta)$ and $w(\theta)$. Without loss of generality we assume that

$$b < u(\theta) < w(\theta) < \min_{1 \le i \le k} \{r_i\} - b, \quad \text{for all } \frac{\pi}{2} \le \theta \le \frac{\pi}{2} \tag{3.39}$$

and that the points **O** and $\mathbf{T}_i$ $(i = 1, \cdots, k)$ cannot be covered by random obscuring disks. The length of the maximum possible radius of an obscuring disk is b.

As in the previous chapter, we order the target points so that $s_1 > s_2 > \cdots > s_k$. We define also

$$\theta_L = s_k - \sin^{-1}\left(\frac{b}{u(s_k)}\right), \tag{3.40}$$

and

$$\theta_U = s_1 + \sin^{-1}\left(\frac{b}{u(s_1)}\right). \tag{3.41}$$

)isks whose centers have orientation angles $\theta < \theta_L$ or $\theta > \theta_U$ cannot intersect any ne of the lines of sight $\overline{OT}_i$ $(i = 1, \cdots, k)$. We focus attention, therefore, only on the ection of $\mathcal{S}$, which contains points with orientation angles θ in $[\theta_L, \theta_U]$. Thus, define he set

$$C = \{(\rho, \theta); u(\theta) \le \rho \le w(\theta); \theta_L \le \theta \le \theta_U\}. \tag{3.42}$$

et $h(\rho, \theta)$ designate a probability density function (p.d.f.), which is concentrated on ', i.e., $h(\rho, \theta) \ge 0$ for all $(\rho, \theta) \in C$, and

$$\int_{\theta_L}^{\theta_U} \int_{u(\theta)}^{w(\theta)} h(\rho, \theta) d\rho\theta = 1. \tag{3.43}$$

his p.d.f. represents the bivariate distribution of the centers of disks over C. In the iform case

$$h(\rho, \theta) = \frac{1}{A\{C\}} \rho, \quad (\rho, \theta) \in C, \tag{3.44}$$

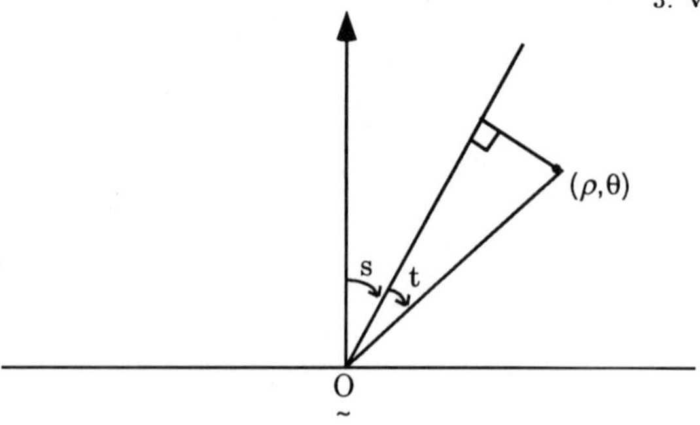

Figure 3.12. The Geometry of Covering Disks

where $A\{C\}$ is the area of C. Let $F(y \mid \rho, \theta)$ be the conditional c.d.f. of the radius of a disk centered at (ρ, θ).

A disk centered at a point (ρ, θ) does not intersect a ray with orientation s only if its radius, Y, is smaller than

$$y(\rho, \theta - s) = \begin{cases} \rho \sin(|\theta - s|), & \text{if } |\theta - s| < \frac{\pi}{2} \\ \rho, & \text{if } |\theta - s| \geq \frac{\pi}{2} \end{cases} \tag{3.45}$$

For angles s, and t, $\theta_L \leq s < s + t \leq \theta_U$, let $K_+(s,t)$ denote the probability that a random disk in C is located between the rays with orientation s and $s + t$, and *does not* intersect the ray with orientation s. In a similar fashion, define $K_-(s,t)$ to be the corresponding probability on the left hand side of the ray with orientation s. Thus, according to (3.45), for every $t > 0$,

$$K_+(s,t) = \int_s^{s+t} \int_{u(\theta)}^{w(\theta)} F(y(\rho, \theta - s) \mid \rho, \theta) h(\rho, \theta) d\rho d\theta \tag{3.46}$$

and

$$K_-(s,t) = \int_{s-t}^{s} \int_{u(\theta)}^{w(\theta)} F(y(\rho, \theta - s \mid \rho, \theta) h(\rho, \theta) d\rho d\theta. \tag{3.47}$$

Due to the properties of the Poisson field, the expected number of disks within C, which are centered between the rays with orientation s and $s \pm t$, and do *not* intersect the ray with orientation s, is $\mu K_\pm(s,t)$. Here μ is the expected number of disk centers which are in C.

Given k points $T_i = (r_i, s_i)$, $i = 1, \cdots, k$, the expected number of disks within C which intersect at least one of the rays $\overline{OT}_i$ $(i = 1, \cdots, k)$ is

$$\Lambda(s_1, \cdots, s_k) = \mu \Bigg[1 - K_+(s_1, \theta_U - s_1) - K_-(s_k, s_k - \theta_L) - \sum_{i=1}^{k-1} \left(K_- \left(s_i, \frac{s_i - s_{i-1}}{2} \right) + K_+ \left(s_{i+1}, \frac{s_i - s_{i+1}}{2} \right) \right) \Bigg], \quad k \geq 2 \tag{3.48}$$

and

$$\Lambda(s_1) = \mu[1 - K_+(s_1, \theta_U - s_1) - K_-(s_1, s_1 - \theta_L)]. \tag{3.49}$$

Finally, the probability that all the k target points will be simultaneously visible from **O** is

$$\psi_k(s_1, \cdots, s_k) = \exp\{-\Lambda(s_1, \cdots, s_k)\}. \tag{3.50}$$

As seen in formulae (3.46) and (3.47), the functions $K_\pm(s,t)$ depend strongly on the geometry of the scattering field. In the following sections we further develop these functions for some special cases.

3.2.2. Standard Poisson Fields with Uniform Distribution of Radii

In the standard case the p.d.f. $h(\rho, \theta)$ is given by Eq. (3.44). The distribution of Y, $F(y)$, is independent of (ρ, θ). Define for $v > 0$

$$\hat{K}(\tau, v) = \int_0^v \rho F(y(\rho, \tau)) d\rho, \tag{3.51}$$

$\hat{K}(\tau, v)$ is an even function of τ, since $y(\rho, \tau)$ depends on τ only through $|\tau|$. Thus,

$$A\{C\} \cdot K_+(s,t) = \int_0^t \hat{K}(\tau, w(\tau + s)) d\tau - \int_0^t \hat{K}(\tau, u(\tau + s)) d\tau. \tag{3.52}$$

We continue the development now, assuming that Y has a uniform distribution over the interval $[a, b]$. Thus, $F(y)$ is given by Eq. (2.44). Substitution in (3.45) yields

$$F(y(\rho,\tau)) = \begin{cases} I\left\{\dfrac{a}{\sin|\tau|} \le \rho < \dfrac{b}{\sin|\tau|}\right\} \cdot \\ \dfrac{\rho\ \sin(\tau) - a}{b - a} + I\left\{\dfrac{b}{\sin|\tau|} \le \rho\right\}, & \text{if } 0 < \tau < \pi/2 \\ F(\rho), & \text{if } \pi/2 \le \tau \end{cases} \tag{3.53}$$

where $I\{A\} = 1$ if $\rho \in A$, and $I\{A\} = 0$ otherwise. Substituting (3.53) in (3.51) we obtain for $\tau > 0$

$$\hat{K}(\tau, v) = \begin{cases} 0, & \text{if } \tau < \sin^{-1}\left(\dfrac{a}{v}\right) \\ \dfrac{1}{b-a}\left[\dfrac{v^3}{3}\sin\ \tau - \dfrac{a}{2}v^2 + \dfrac{1}{6}\dfrac{a^3}{\sin^2\ \tau}\right], & \text{if } \sin^{-1}\left(\dfrac{a}{v}\right) \le \tau < \sin^{-1}\left(\dfrac{b}{v}\right) \\ \dfrac{v^2}{2} - \dfrac{1}{6} \cdot \dfrac{a^2 + ab + b^2}{\sin^2\ \tau}, & \text{if } \sin^{-1}\left(\dfrac{b}{v}\right) \le \tau \le \dfrac{\pi}{s} \\ \dfrac{v^2}{2} - \dfrac{1}{6}(a^2 + ab + b^2), & \text{if } \dfrac{\pi}{2} < \tau \end{cases} \tag{3.54}$$

In the following sections we consider the effect of the structure of C on $K_\pm(s,t)$.

3.2.2.1. Annular Scattering Region

If the scattering region S is annular the boundaries of C are two parallel circular arcs

$$u(\theta) = u, \quad \text{all} \quad \theta_L \le \theta \le \theta_U$$
$$w(\theta) = w, \quad \text{all} \quad \theta_L \le \theta \le \theta_U,$$

where $b < u < w$.

Moreover, in this case $K_+(s,t) = K_-(s,t)$. Indeed,

$$\int_{-t}^{0} \hat{K}(\tau, w)d\tau = \int_{0}^{t} \hat{K}(-\tau, w)d\tau = \int_{0}^{t} \hat{K}(\tau, w)d\tau,$$

since $\hat{K}(\tau, w)$ is an even function of τ. Accordingly (3.52) implies that in the case of an annular region $K_+(s,t) = K_-(s,t) \equiv K(s,t)$. Furthermore, this function does not depend on s. Finally, let

$$K^*(t,v) = \int_0^t \hat{K}(\tau, v)d\tau,$$

then, integrating (3.54) we obtain for $v \ge b$

$$K^*(t,v) = \begin{cases} 0, & t < \sin^{-1}\left(\frac{a}{v}\right) \\ K_1(t,v), & \sin^{-1}\left(\frac{a}{v}\right) \le t < \sin^{-1}\left(\frac{b}{v}\right) \\ K_2(t,v), & \sin^{-1}\left(\frac{b}{v}\right) \le t < \frac{\pi}{2} \\ K_3(t,v), & \frac{\pi}{2} \le t \end{cases} \tag{3.55}$$

where

$$K_1(t,v) = \frac{1}{b-a}\left[\frac{v^3}{3}\left(\left(1-\left(\frac{a}{v}\right)^2\right)^{1/2} - \cos(t)\right) - \frac{a}{2}v^2\left(t - \sin^{-1}\left(\frac{a}{v}\right)\right) + \frac{a^2}{6}\left(v\left(1-\left(\frac{a}{v}\right)^2\right)^{1/2} - a\cot(t)\right)\right], \tag{3.56}$$

$$K_2(t,v) = K_1\left(\sin^{-1}\left(\frac{b}{v}\right), v\right) + \frac{v^2}{2}\left(t - \sin^{-1}\left(\frac{b}{v}\right)\right) + \frac{a^2+ab+b^2}{6}\left[\cot(t) - \frac{v}{b}\left(1-\left(\frac{b}{v}\right)^2\right)^{1/2}\right], \tag{3.57}$$

and

$$K_3(t,v) = K_2\left(\frac{\pi}{2}, v\right) + \left(t - \frac{\pi}{2}\right)\left(\frac{v^2}{2} - \frac{a^2+ab+b^2}{6}\right). \tag{3.58}$$

Notice that $K(s,t) = K^*(t,w) - K^*(t,u)$. The simultaneous visibility probability attains in the annular case the formula, with $\lambda = \mu/A\{C\}$ in units of $[1/m^2]$,

$$\psi_k(s_1,\cdots,s_k) = \exp\Big\{ -\lambda\Big[A\{C\} - K^*(\theta_U - s_1, w) + K^*(\theta_U - s_1, u) - K^*(s_k - \theta_L, w) + K^*(s_k - \theta_L, u) - 2\sum_{i=1}^{k-1}\Big(K^*\Big(\frac{s_i - s_{i-1}}{2}, w\Big) - K^*\Big(\frac{s_i - s_{i-1}}{2}, u\Big)\Big)\Big]\Big\},$$

where

$$A\{C\} = \frac{1}{2}(w^2 - u^2)(\theta_U - \theta_L)$$

is the area of the annular region. The program VPANN computes $\psi_k(s_1,\cdots,s_k)$.

Example 3.7. Consider an annular region with $u = 50$[m] and $w = 75$[m]. The target points have orientation angles of 45°, 44°, 10°, −5°, −15°. Suppose that the distribution of the random radius of a disk is uniform on [0.5,1]. Program VPANN yields the following visibility probabilities.

Table 3.2. Simultaneous Visibility Probabilities Annual Regions

j	$\lambda = .001[1/\text{m}^2]$	$\lambda = .01[1/\text{m}^2]$
1	.9630	.6856
2	.9373	.5234
3	.9042	.3651
4	.8696	.2472
5	.8363	.1674

The values in the j-th row ($j = 1,\cdots,5$) of the table are the simultaneous visibility probability of the first j target points. ∎

3.2.2.2. Trapezoidal Scattering Regions

In a trapezoidal region $u(\theta)$ and $w(\theta)$ are horizontal parallel straight lines of distance u and w from **O**, respectively. The formulae of $u(\theta)$ and $w(\theta)$ are

$$u(\theta) = \frac{u}{\cos\ \theta}, \qquad \theta_L \leq \theta \leq \theta_U$$

and

$$w(\theta) = \frac{2}{\cos\ \theta}, \qquad \theta_L \leq \theta \leq \theta_U.$$

According to (3.52),

$$A\{C\}K_+(s,t) = \int_0^t \left[\hat{K}\left(\tau, \frac{w}{\cos(\tau + s)}\right) - \hat{K}\left(\tau, \frac{u}{\cos(\tau + s)}\right)\right] d\tau.$$

Moreover, since $\hat{K}\left(\tau, \frac{c}{\cos(\tau+s)}\right) = \hat{K}\left(-\tau, \frac{c}{\cos(-\tau-s)}\right)$, for any $c > 0$, we obtain the relationship $K_-(s,t) = K_+(-s,t)$, for all s and t. We develop now an explicit formula for $K_+(s,t)$ for the case where the radii of obscuring disks are uniformly distributed on the interval (a,b). In order to write $\hat{K}\left(\tau, \frac{v}{\cos(\tau+s)}\right)$ as an explicit function of τ, for fixed values of v and s, $v > 0$, we consider first the root $\tau \equiv \eta(x,s)$ of the equation

$$\tau = \sin^{-1}(x \ \cos(\tau+s)), \quad 0 < \tau < \pi \ \text{ for } \ 0 < x < 1, \ \frac{-\pi}{2} < s < \frac{\pi}{2}. \tag{3.61}$$

Notice that $\sin^{-1}(x \ \cos(\tau+s))$ is a strictly decreasing function of τ, which is positive at $\tau = 0$ and zero at $\tau = \frac{\pi}{2} - s$. Hence, there exists a unique root $\tau \equiv \eta(x,s)$ of Eq. (3.61), given by

$$\begin{aligned} \eta(x,s) &= \tan^{-1}\left(\frac{x \ \cos \ s}{1 + x \ \sin \ s}\right) \\ &= \tan^{-1}\left(\tan \ s + \frac{x}{\cos \ s}\right) - s. \end{aligned} \tag{3.62}$$

The function $\eta(x,s)$ is increasing in x, for a fixed s, and $\eta(0,s) = 0$.

Let $K^*(\tau,s,v) = \hat{K}\left(\tau, \frac{v}{\cos(\tau+s)}\right)$. According to Eq. (3.54),

$$K^*(\tau,s,v) = \begin{cases} 0, & \tau < \eta\left(\frac{a}{v}, s\right) \\ \frac{1}{6(b-a)}\left[2v^3 \frac{\sin \ \tau}{\cos^3(\tau+s)} \right. & \\ \left. -3a\frac{v^2}{\cos^2(\tau+s)} + \frac{a^3}{\sin^2 \ \tau}\right], & \eta\left(\frac{a}{v}, s\right) \le \tau \le \eta\left(\frac{b}{v}, s\right) \\ \frac{v^2}{2 \ \cos^2(\tau+s)} - \frac{a^2+ab+b^2}{6 \ \sin^2 \ \tau}, & \eta\left(\frac{b}{v}, s\right) \le \tau \le \frac{\pi}{2} \\ \frac{v^2}{2 \ \cos^2(\tau+s)} - \frac{a^2+ab+b^2}{6}, & \frac{\pi}{2} < \tau. \end{cases} \tag{3.63}$$

Define the function $\tilde{K}(t,s,v) = \int_0^t K^*(\tau,s,v)d\tau$. According to (3.62) and (3.63),

$$\tilde{K}(t,s,v) = \begin{cases} 0, & \text{if } t < \eta\left(\frac{a}{v}, s\right) \\ K^{(1)}(t,s,v), & \text{if } \eta\left(\frac{a}{v}, s\right) \le t < \eta\left(\frac{b}{v}, s\right) \\ K^{(2)}(t,s,v), & \text{if } \eta\left(\frac{b}{v}, s\right) \le t < \frac{\pi}{2} \\ K^{(3)}(t,s,v), & \text{if } \frac{\pi}{s} \le t, \end{cases}$$

where

$$\begin{aligned} K^{(1)}(t,s,v) = \frac{1}{6(b-a)} \Bigg\{ & \frac{v^3}{\cos(s)} \left[\frac{\sin^2(t)}{\cos^2(s+t)} \right. \\ & \left. - \left(\frac{a}{v}\right)^2 \cos^2(s) \frac{1 + (\tan\ s + \frac{a}{v\ \cos(s)})^2}{1 + 2\frac{a}{v}\sin(s) + \left(\frac{a}{v}\right)^2} \right] \\ & - 3av^2 \left[\tan(t+s) - \tan(s) - \frac{a}{v\ \cos(s)} \right] \\ + \quad & a^3 \left[\frac{a}{v\ \cos(s)} + \tan(s) - \cot(t) \right] \Bigg\}, \end{aligned} \tag{3.64}$$

$$\begin{aligned} K^{(2)}(t,s,v) = K^{(1)} & \left(\eta\left(\frac{b}{v}, s\right), s, v \right) \\ & + \frac{v^2}{2} \left[\tan(t+s) - \tan(s) - \frac{b}{v\ \cos(s)} \right] \\ & - \frac{a^2 + ab + b^2}{6} \left[\frac{b}{v\ \cos(s)} + \tan(s) - \cot(t) \right], \end{aligned} \tag{3.65}$$

ıd

$$K^{(3)}(t,s,v) = K^{(2)}\left(\frac{\pi}{2}, s, v\right) + \frac{v^2}{2}[\tan(t+s) + \cot(s)] - \frac{a^2+ab+b^2}{6}. \tag{3.66}$$

Finally, $A\{C\}K_+(s,t) = \tilde{K}(s,t,w) - \tilde{K}(s,t,u)$.

ːample 3.8. We illustrate the above results numerically with $u = 50$[m], $w =$ [m]. We consider 5 target points with orientations 70°, 60°, 50°, 45° and 40°. The tribution of disk radius is uniform on (a,b), with $a = 0$ and $b = 1$[m]. The program ·TRA yields the following results.

Table 3.3. Simultaneous Visibilities in a Trapezoidal Region.

k	$\lambda = .001[1/\text{m}^2]$	$\lambda = .01[1/\text{m}^2]$
1	.9295	.4815
2	.8842	.2920
3	.8505	.1979
4	.8209	.1390
5	.7946	.1003

The tabulated value for a given k is the simultaneous visibility probability for the first k points. Verify that program SIMVP yields the same results. ■

3.3. An Alternative Geometric-Analytic Method

In the present section we construct the K-functions by a combination of the geometric and analytic methods, restricting attention to the trapezoidal regions, defined in Section 3.2.2.2. We start with the development of a method for the determination of $K_+(s,t)$. Recall that a Poisson field has the property that, given the number N, of disks scattered over the region C, and given any partition of C to sets $B_1, B_2, \cdots, B_m$, the number of disks having centers within these sets, $J_1, J_2, \cdots, J_m$ have a joint m-nomial distribution, with parameters $(N, \theta_1, \cdots, \theta_m)$, where

$$\theta_j = \iint_{B_j} h(x,y)dxdy/H\{C\}, \quad j = 1, \cdots, m. \tag{3.67}$$

The function $h(x,y)$ is a scattering density of the centers of disks in terms of Cartesian coordinates, and

$$H\{C\} = \iint_C h(x,y)dxdy.$$

The function $K_+(s,t)$ is the probability that a random disk, centered between th rays $\mathbb{R}_s$ and $\mathbb{R}_{s+t}$, with orientation angles s and $s+t$, does not intersect the ray $\mathbb{R}_s$. Let $K_+(s,t,r)$ be the **conditional** probability that a random disk, having a radius $a \leq r \leq b$, and centered between the rays $\mathbb{R}_s$ and $\mathbb{R}_{s+t}$, does not intersect $\mathbb{R}_s$. This equivalent to the probability that the disk center belongs to the set $B_+(r)$ (see Figu 3.13).

Assuming that the radii of random disks are uniformly distributed over $[a,b]$, ind pendently of their location,

$$\begin{aligned} K_+(s,t) &= \frac{1}{b-a}\int_a^b K_+(s,t,r)dr \\ &= \frac{1}{b-a}\int_a^b \Pr\{(X,Y) \in B_+(r)\}dr, \end{aligned} \tag{3.6}$$

where

$$\Pr\{(X,Y) \in B_+(r)\} = \frac{1}{H\{C\}}\iint_{B_+(r)} h(x,y)dxdy. \tag{3.6}$$

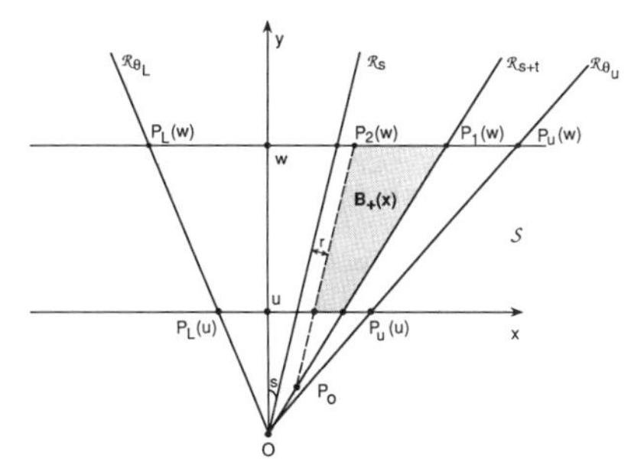

Figure 3.13. The Geometry of $B_{+}(r)$

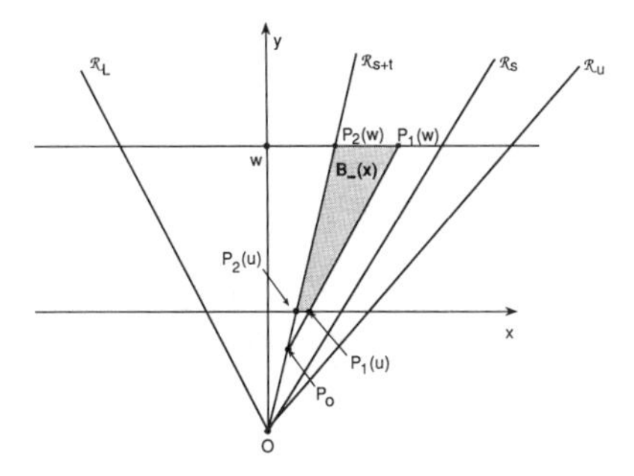

Figure 3.14. The Geometry of $B_{-}(r)$.

Let $\mathbf{P}_1(v)$ be the point of intersection of $\mathbb{R}_{s+t}$ with the horizontal line of distance from the origin $\mathbf{O}$. Let $\mathbf{P}_2(v)$ be a point of intersection a line parallel to $\mathbb{R}_s$, to its ight, at distance r from it, and a horizontal line of distance v from the origin. Let $_0$ be the point of intersection of the line through $\mathbf{P}_2(u)$ and $\mathbf{P}_2(w)$ and $\mathbb{R}_{s+t}$. The et $B_{+}(r)$ is **empty** if the y-coordinate of $\mathbf{P}_0$ is greater than w. The set $B_{+}(r)$ is the iangle $\Delta(\mathbf{P}_0, \mathbf{P}_1(w), \mathbf{P}_2(w))$, if the y-coordinate of $\mathbf{P}_0$ is greater than u but smaller ıan w. $B_{+}(r)$ is the trapezoid $(\mathbf{P}_2(u), \mathbf{P}_1(u), \mathbf{P}_1(w), \mathbf{P}_2(w))$ if the y-coordinate of $\mathbf{P}_0$ is naller than u. This trapezoid (see Figure 3.13) is the difference $\Delta(\mathbf{P}_0, \mathbf{P}_1(w), \mathbf{P}_2(w)) -$ $(\mathbf{P}_0, \mathbf{P}_1(u), \mathbf{P}_2(u))$. Accordingly, the integral in (3.69), when $B_{+}(r)$ is not empty, is e probability that a random center (X, Y) belongs to a triangle, or is the difference etween two such triangular probabilities. Further details on the computations of such iangular probabilities, in the bivariate normal case, will be given in Section 3.3.1.

The function $K_{-}(s,t)$ can be obtained by the same method. Define

$$K_{-}(s,t) = \frac{1}{b-a} \int_a^v \Pr\{(X,Y) \in B_{-}(r)\} dr. \tag{3.70}$$

ıere $B_{-}(r)$ is again a triangle or a trapezoid (see Figure 3.14).

Let B be any set in the (x,y) plane, and let $(B)^{-}$ denote the set obtained by the ınsformation $x \to -x$ and $y \to y$ (mirror reflection with respect to the y-axis). Simrly, let $\mathbf{P}^{-}$ be the reflection of $\mathbf{P}$ obtained by this transformation. Then, if $B_{-}(r)$

is non-empty, we can consider the reflected set $(B_-(r))^-$, which is either the triangle $\Delta(\mathbf{P}_0^-, \mathbf{P}_2^-(w), \mathbf{P}_1^-(w))$, when $u \leq y_0 < w$, or the trapezoid $(\mathbf{P}_1^-(u), \mathbf{P}_2^-(u), \mathbf{P}_2^-(w), \mathbf{P}_1^-(w))$

$$\begin{aligned} \Pr\{(X,Y) \in B_-(r)\} &= \frac{1}{H\{C\}} \iint\limits_{B_-(r)} h(x,y)dxdy \\ &= \frac{1}{H\{C\}} \iint\limits_{(B_-(r))^-} h^-(x,y)dxdy, \end{aligned} \tag{3.71}$$

where $h^-(x,y)$ is the p.d.f. of $(-X,Y)$.

3.3.1. Computing the Probability of $B_+(r)$ in the Bivariate Normal Case

In the present section we assume that $h(x,y)$ is the p.d.f. of a bivariate normal distribution (Eq. (1.61)). As seen in Figure 3.13, the set $B_+(r)$ is the difference of two triangles $\Delta(\mathbf{P}_0, \mathbf{P}_1(w), \mathbf{P}_2(w))$ and $\Delta(\mathbf{P}_0, \mathbf{P}_1(u), \mathbf{P}_2(u))$. The special feature of these two triangles is that the side connecting $\mathbf{P}_1(w)$ and $\mathbf{P}_2(w)$ is a straight line parallel to the x-axis.

Let (x_0, y_0) be the coordinates of $\mathbf{P}_0$, (x_1, y_1) the coordinates of $\mathbf{P}_1(w)$ and (x_2, y_2) the coordinates of $\mathbf{P}_2(w)$. It is easy to check that

$$\begin{aligned} x_1 &= w \tan(s+t) \\ y_1 &= w, \\ x_2 &= (w \sin(s) + r)/\cos(s), \\ y_2 &= w, \end{aligned} \tag{3.72}$$

and

$$\begin{aligned} x_0 &= r\frac{\sin(s+t)}{\Delta(s,t)}, \\ y_0 &= r\frac{\cos(s+t)}{\Delta(s,t)}, \end{aligned} \tag{3.73}$$

where

$$\Delta(s,t) = \cos(s)\sin(s+t) - \sin(s)\cos(s+t). \tag{3.74}$$

Let $T(x_0, y_0, x_1, y_1, x_2, y_2)$ denote the bivariate normal probability of the triang[le] $\Delta(x_0, y_0, x_1, y_1, x_2, y_2)$. We denote by $\mathbf{P}_0$ the vertex having the smallest y-coordinat[e] and by $\mathbf{P}_1$ and $\mathbf{P}_2$ the other two vertices, following the sides of the triangle in a count[er]clockwise direction. We have five distinguishable cases, according to the relative positi[on] of the coordinates x_0, x_1 and x_2. We develop here explicit formulae only for case namely $x_2 < x_1 < x_0$ (see Figure 3.15).

Consider a bivariate normal distribution with mean vector (ξ, η) and covariate matr[ix] $V = \begin{bmatrix} \sigma_x^2 & \rho\sigma_x\sigma_y \\ \rho\sigma_x\sigma_y & \sigma_y^2 \end{bmatrix}$. Let $\beta = \rho\sigma_y/\sigma_x$. The conditional distribution of Y given $X =$

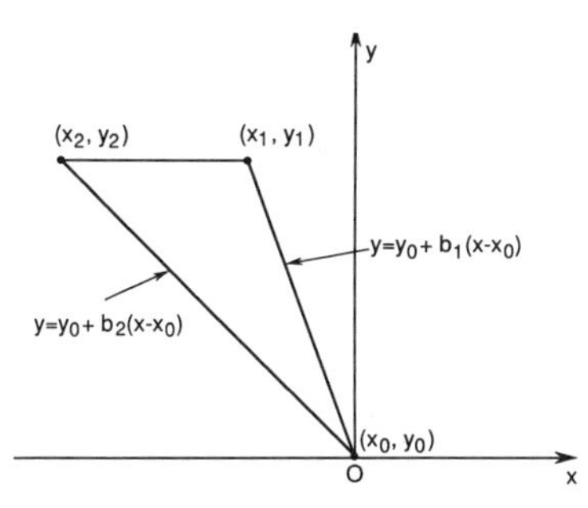

Figure 3.15. The Geometry in Case I

is normal, with mean $\eta + \beta(x-\xi)$ and variance $\sigma_y^2(1-\rho^2)$. Thus, as shown in Section 2.6.2,

$$\begin{aligned} T(x_0, y_0, x_1, w, x_2, w) = & \frac{1}{\sigma_x}\int_{x_2}^{x_1} \phi\left(\frac{x}{\sigma_x}\right)\left[\Phi\left(\frac{w-\eta-\beta(x-\xi)}{\sigma_y(1-\rho^2)^{1/2}}\right)\right. \\ & \left. - \Phi\left(\frac{y_0+b_2(x-x_0)-\eta-\beta(x-\xi)}{\sigma_y(1-\rho^2)^{1/2}}\right)\right]dx \\ & + \frac{1}{\sigma_x}\int_{x_1}^{x_0} \phi\left(\frac{x}{\sigma_x}\right)\left[\Phi\left(\frac{y_0+b_1(x-x_0)-\eta-\beta(x-\xi)}{\sigma_y(1-\rho^2)^{1/2}}\right)\right. \\ & \left. - \Phi\left(\frac{y_0+b_2(x-x_0)-\eta-\beta(x-\xi)}{\sigma_y(1-\rho^2)^{1/2}}\right)\right]dx. \end{aligned} \tag{3.75}$$

otice that $y_0 + b_1(x - x_0)$ and $y_0 + b_2(x - x_0)$ are the formulae of the straight lines onnection $\mathbf{P}_0$ with $\mathbf{P}_1$ and $\mathbf{P}_0$ with $\mathbf{P}_2$, respectively (see Figure 3.15). The integrals in 3.75) can be evaluated numerically by proper integration routines. We provide a pro-am VPNORM, (see programs for Chapter 5) which computes visibility probabilities r trapezoidal regions, with bivariate normal scattering. Notice that in this case,

$$\begin{aligned} H\{C\} = & T(0, 0, x_U(w), w, x_L(w), w) \\ & - T(0, 0, x_U(u), u, x_L(u), u) \end{aligned} \tag{3.76}$$

nere $(x_U(w), w)$ are the coordinates of $\mathbf{P}_U(w)$, $(x_L(w), w)$ are those of $\mathbf{P}_L(w)$, etc. e Figure 3.13).

xample 3.9. We compute here the simultaneous visibility probability of $n = 5$ get points having orientation angles of 45°, 44°, 5°, −15°, and 15°. The scattering gion S is between the horizontal lines of distances $u = 50$[m] and $w = 75$[m]. The ttering density of centers of disks is the bivariate normal with means $\xi = 0$, $\eta = 50$[m], $= \sigma_y = 30$[m] and $\rho = .5$. The intensity is $\mu = 25$ trees, over the region C. The lii of the trees are uniformly distributed between .5[m] and 1[m]. The simultaneous ibility probabilities, computed according to program "VPNORM" are

Table 3.4. Simultaneous Visibility Probabilities

i	VP
1	.9200
2	.8631
3	.6211
4	.4493
5	.3354

The i-th row in the above table gives the simultaneous visibility probability of the first i points. ■

3.4. The Visibility of Windows

It is often the case that one needs a minimal neighborhood of length l_0 around a point **T** to be completely visible. In Eqs. (3.5) and (3.16) we presented the probabilities that linear "windows" of length L are completely visible around a single target point. In the present section we show how the simultaneous visibility probability of several windows can be determined. We do it by modifying the K-functions in the following manner.

If s is the orientation angle of a target point of distance $\rho(s)$ from the origin, **O**, compute

$$\alpha_l(s) = \sin^{-1}(l_0/2\rho(s)) \tag{3.77}$$

and change $K_+(s,t)$ to $K_+(s+\alpha_l(s), t_l(s))$, where

$$t_l(s) = (t - \alpha_l(s))^+, \tag{3.78}$$

and where $(a)^+ = \max(a, 0)$. Recall that $K_+(\cdot, 0) \equiv 0$. Similarly, $K_-(s,t)$ is changed to $K_-(s - \alpha_l(s), t_l(s))$. If the target point is on a star-shaped curve $\mathcal{C}$, we determine the angle $\alpha_l^+(s)$ so that the length of the portion o the curve $\mathcal{C}$, between the rays $\mathbb{R}_s$ and $\mathbb{R}_{s+\alpha_l(s)}$ is $l/2$. For example, if $\mathcal{C}$ is a circular path of radius c, then $\alpha_l^+(s) = l/2c$ for all s. If, on the other hand, $\mathcal{C}$ is a horizontal line, of distance v from **O**, then $\alpha_l^+(s) = \tan^{-1}\left(\tan(s) + \frac{l}{2v}\right) - s$. Similarly, define the angle $\alpha_l^-(s)$ so that the length of $\mathcal{C}$ between $\mathbb{R}_{s-\alpha_l^-(s)}$ and $\mathbb{R}_s$ is $l/2$. Again, in the case of a circular curve $\mathcal{C}$, $\alpha_l^-(s) = \alpha_l^+(s) = \frac{l}{2c}$. On the other hand, if $\mathcal{C}$ is a horizontal line, then $\alpha_l^-(s) = s - \tan^{-1}\left(\tan(s) - \frac{l}{2v}\right) = \alpha_l^+(-s)$.

Example 3.10. We compute the simultaneous visibility probabilities of 5 target points with windows of size l[m] when the region $\mathcal{C}$ is annular, with $u = 50$[m], $w = 75$[m] and the target are along a circular curve with radius of $r = 100$[m]. The orientation angles of the targets are: $s = 45°$, $44°$, $39°$, $0°$, $-1°$. The scattering of disks is according to a homogeneous Poisson process with $\lambda = .001[1/\text{m}^2]$ and $\lambda = 0.01[1/\text{m}^2]$, $a = b = 1.0$[m]. The results in the following table were computed with program VPANNW

Table 3.5. Visibility Probabilities With Windows

	λ	=	0.001	λ	=	0.01
i	$l = 0$	$l = 1$	$l = 2$	$l = 0$	$l = 1$	$l = 2$
1	0.963	0.949	0.937	0.687	0.590	0.522
2	0.938	0.923	0.912	0.525	0.449	0.397
3	0.903	0.876	0.851	0.360	0.264	0.200
4	0.870	0.830	0.795	0.248	0.155	0.100
5	0.847	0.808	0.776	0.189	0.119	0.079

The values at the ith row are the simultaneous visibility probabilities of the first i target points neighborhood (windows). ■

Program MOMTVPW can be used to compute visibility probabilities with windows in case of trapezoidal regions.

4

Visibility Probabilities II

In the present chapter we consider problems associated with the visibility of targets from several observation points, subject to interference by a random field of obscuring elements, and visibility probabilities in three dimensional spaces. This chapter presents extensions of the results of Chapter 3. We consider a random field of obscuring elements, which are centered in a region located between the observation points and the targets. The targets are stationary points in the plane, or in space.

4.1. The Multi-Observer Multi-Target Shadowing Model and Simultaneous Visibility Probabilities

In the present section we model the field of obscuring elements, relative to the n observation points and the m target points.

Let $\mathcal{S}$ be a strip in the plane, bounded between two parallel lines, $\mathcal{U}$, $\mathcal{W}$. Without loss of generality, assume that $\mathcal{U}$ and $\mathcal{W}$ are parallel to the x-axis, and pass through the points $(0, U)$ and $(0, W)$, respectively. A finite number of disks are randomly centered in $\mathcal{S}$, according to a Poisson process with intensity function $\lambda h(x, y)$. We further assume that radii of disks centered in $\mathcal{S}$ are independent random variables, having distribution which may depend on their locations.

Denote by $F(r \mid x, y)$ the conditional c.d.f. of the radius of a disk centered at (x, y) It is assumed that all these conditional distributions are concentrated on the interva $[0, b]$.

Let $\mathbf{O}_1, \cdots, \mathbf{O}_\nu$ be the specified observation points, and let $\mathbf{T}_1, \cdots, \mathbf{T}_m$ be the spec ified target points. We assume that all the observation points are on one side of $\mathcal{S}$ an all the target points are on the other side of it. We also assume that $\mathbf{O}_i$ $(i = 1, \cdots, \nu$ and $\mathbf{T}_j$ $(j = 1, \cdots, m)$ are not covered by random disks. That is, all the ordinates o $\mathbf{O}_i$ are smaller than $U - b$ and all the ordinates of $\mathbf{T}_j$ are greater than $W + b$.

We say that a target point $\mathbf{T}_j$ is visible from an observation point $\mathbf{O}_i$, if the line-o sight, $\mathcal{L}_{ij}$, connecting $\mathbf{O}_i$ with $\mathbf{T}_j$ is not intersected by random disks in $\mathcal{S}$.

In Figure 4.1 we illustrate the geometrical structure for the case of two observatio points and three targets. In this case there are six possible lines of sight. Generally the are $N = \nu \cdot m$ lines of sight. In the previous section we denoted the line of sight fro $\mathbf{O}_i$ to $\mathbf{T}_j$ by $\mathcal{L}_{ij}$. We now order these lines according to the index $n(i, j) = m^{i-1} +$ Accordingly, the line of sight from $\mathbf{O}_2$ to T_2 in Figure 4.1 is indexed as $\mathcal{L}_5$.

The obscuring disks are centered within the strip $\mathcal{S}$, between the two parallel lin $\mathcal{U}$ and $\mathcal{W}$. We see in Figure 4.1 that the six lines $\mathcal{L}_1, \cdots, \mathcal{L}_6$ intersect within $\mathcal{S}$ ar fo different points. Due to these intersections the relative positions of the lines $\mathcal{L}_\nu$, wi respect to each other change, and the formulae of joint visibility probabilities deriv earlier, may not be directly applicable. For this reason we develop here a methodolo which is appropriate for this situation. This methodology is based on partitioning t scattering region $\mathcal{S}$ to subregions in which the lines of sight do not intersect. Mo specifically, suppose that the N lines of sight intersect at K^* points within $\mathcal{S}$ $(K^* \geq$ We draw lines parallel to the x-axis, which pass through these intersection points. Th

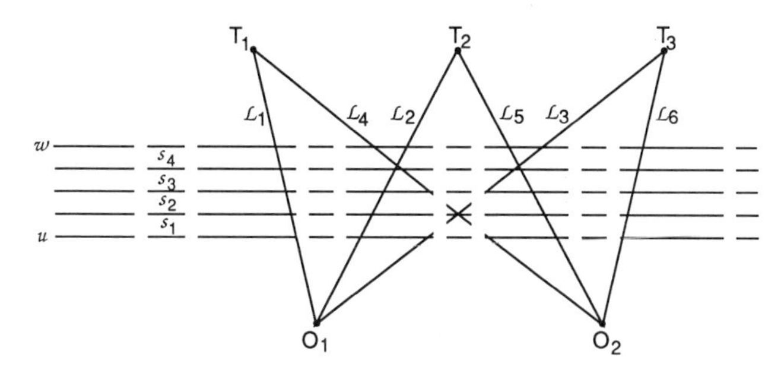

Figure 4.1. Lines of Sight From Several Observation Points

lines partition $\mathcal{S}$ into K ($K \le K^* + 1$) parallel strips, $\mathcal{S}_1, \mathcal{S}_2, \cdots, \mathcal{S}_K$. The k-th strip, $\mathcal{S}_k$, is bounded between the two parallel lines $\mathcal{U}_{k-1}$ and $\mathcal{U}_k$, where the line $\mathcal{U}_k$ intersects the y-axis at the point $(0, U_k)$, $k = 1, \cdots, K$; and where $\mathcal{U}_0 \equiv \mathcal{U}$, $\mathcal{U}_K = \mathcal{W}$.

Consider N lines of sight, $\mathcal{L}_1, \cdots, \mathcal{L}_N$. Let $M_k^{(N)}$ denote the expected number of disks centered in $\mathcal{S}_k$, $k = 1, \cdots, K$, which intersect at least one of the lines $\mathcal{L}_n$ ($n = 1, \cdots, N$). The joint visibility probability of the N lines is

$$\psi(\mathcal{L}_1, \cdots, \mathcal{L}_N) = \exp\left\{ -\sum_{k=1}^{K} M_k^{(N)} \right\}. \tag{4.1}$$

Let $m_k(n, n')$ denote the expected number of disks centered between $\mathcal{L}_n$ and a line $\mathcal{L}_{n'}$ to its right in $\mathcal{S}_k$, which intersect either $\mathcal{L}_n$ or $\mathcal{L}_{n'}$. $m_k(0, n')$ is the expected number of disks to the left of $\mathcal{L}_{n'}$ in $\mathcal{S}_k$, which intersect it. Similarly, let $m_k(n, N+1)$ denote the expected number of disks to the right of $\mathcal{L}_n$, in $\mathcal{S}_k$, which intersect it. Let $\nu_k^+(n)$ denote the index of the line which is in $\mathcal{S}_k$ closest to $\mathcal{L}_n$ on the right. Notice that for a given index n ($n = 1, \cdots, N-1$), $\nu_k^+(n)$ may change values at different substrips $\mathcal{S}_k$. In order to show the dependence of $\nu_k^+(n)$ on k consider the example illustrated in Figure 4.1. The strip $\mathcal{S}$ is partitioned into 4 substrips. In the following table we present the indexes $\nu_k^+(n)$ in these four substrips.

Table 4.1. Indexes $\nu_k^+(n)$ for $n = 1, \cdots, 5$; $k = 1, \cdots, 4$

$k = 1$		$k = 2$		$k = 3$		$k = 4$	
n	$\nu_k^+(n)$	n	$\nu_k^+(n)$	n	$\nu_k^+(n)$	n	$\nu_k^+(n)$
1	2	1	2	1	4	1	4
2	3	2	4	2	3	2	5
3	4	3	5	3	5	3	6
4	5	4	3	4	2	4	2
5	6	5	6	5	6	5	3

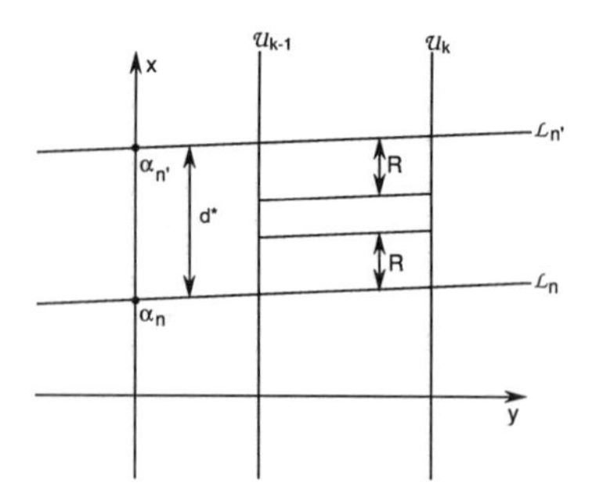

Figure 4.2. The Geometry with Parallel Lines

It follows that, for each $k = 1, \cdots, K$,

$$M_k^{(N)} = m_k(0,1) + \sum_{n=1}^{N-1} m_k(n, \nu_k^+(n)) + m_k(N, N+1) \tag{4.2}$$

In the following section we derive formulae for $m_k(n, n')$.

4.2. General Formulae for $m_k(n, n')$ for the Standard Poisson Field

The equation of $\mathcal{L}_n$ can be expressed as

$$\mathcal{L}_n : x_n(y) = \alpha_n + \beta_n y.$$

Given two lines, $\mathcal{L}_n$ and $\mathcal{L}_{n'}$ we distinguish between two major cases for deriving th formula of $m_k(n, n')$, according to whether the lines are parallel ($\beta_n = \beta_{n'}$) or no ($\beta_n \neq \beta_{n'}$).

Case I: $\beta_n = \beta_{n'}$.

The line $\mathcal{L}_{n'}$ is to the right of $\mathcal{L}_n$ if $\alpha_{n'} > \alpha_n$ (see Figure 4.2). The distance betwee $\mathcal{L}_n$ and $\mathcal{L}_{n'}$ is

$$d^*_{nn'} = (\alpha_{n'} - \alpha_n)/\gamma_n, \qquad (4.3$$

where $\gamma_n = (1 + \beta_n^2)^{1/2}$.

A disk of radius R centered in $\mathcal{S}_k$ between $\mathcal{L}_n$ and $\mathcal{L}_{n'}$ intersects either line if i distance from them is less than R. Let $m_k(n, n'; R)$ denote the expected number disks of radius R, centered at $\mathcal{S}_k$ between $\mathcal{L}_n$ and $\mathcal{L}_{n'}$, which intersect either one of tl lines. By computing areas of parallelograms (see Figure 4.2) one can verify that

$$m_k(n, n'; R) = \begin{cases} \lambda 2R\gamma_n(U_k - U_{k-1}), & \text{if } R < d^*/2 \\ \\ \lambda d^*\gamma_n(U_k - U_{k-1}), & \text{if } R \geq d^*/2. \end{cases}$$

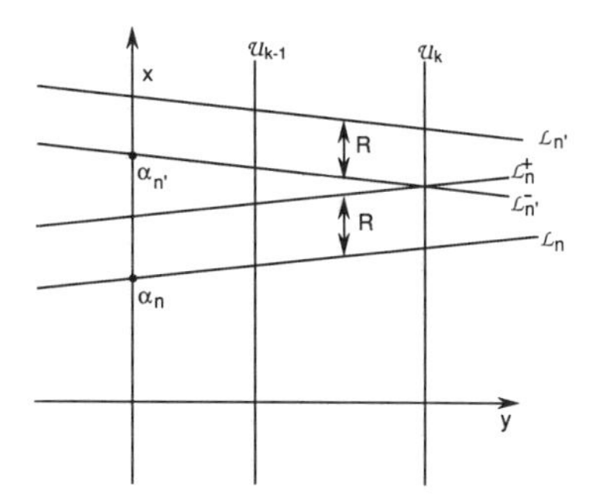

Figure 4.3. The Geometry of Unparallel Lines

Let $F(r)$ denote the c.d.f. of a disk radius and $f(r)$ its p.d.f. The formula of $m_k(n, n')$ when $\beta_n = \beta_{n'}$ is then

$$m_k(n, n') = \lambda \gamma_n (U_k - U_{k-1}) \left[d^* \left(1 - F\left(\frac{d^*}{2}\right)\right) + 2\int_0^{d^*/2} r f(r) dr \right]. \tag{4.5}$$

Case II: $\beta_n \neq \beta_{n'}$.

We derive first the formula of $m_k(n, n')$ under the assumption that $\beta_n > \beta_{n'}$. Let

$$\Delta_{nn'}(y) = x_{n'}(y) - x_n(y) = (\alpha_{n'} - \alpha_n) + (\beta_{n'} - \beta_n)y.$$

Obviously, since $\beta_n > \beta_{n'}$, $\Delta_{nn'}(U_{k-1}) > \Delta_{nn'}(U_k)$. Let $\xi_{nn'}(y) = \Delta_{nn'}(y)/(\gamma_n + \gamma_{n'})$. Consider the lines

$$\mathcal{L}_{n,R}^+ : x_{n,R}^+ = \alpha_n + \gamma_n R + \beta_n y$$
$$\mathcal{L}_{n,R}^- : x_{n',R}^- = \alpha_{n'} - \gamma_{n'} R + \beta_{n'} y.$$

A disk of radius R intersects $\mathcal{L}_n$ if it is centered between $\mathcal{L}_n$ and $\mathcal{L}_{n,R}^+$. Such a disk ntersects $\mathcal{L}_{n'}$ if it is centered between $\mathcal{L}_{n',R}^-$ and $\mathcal{L}_{n'}$.

If $R < \xi_{nn'}(U_k)$ the lines $\mathcal{L}_{n,R}^+$ and $\mathcal{L}_{n,R}^-$ intersect at $y^* > u_k$, and the expected umber of disks, of radius R, which intersect either $\mathcal{L}_n$ or $\mathcal{L}_{n'}$ is

$$m_k(n, n'; R) = \lambda(\gamma_n + \gamma_{n'})(U_k - U_{k-1})R. \tag{4.6}$$

f $\xi_{nn'}(U_k) < R < \xi_{nn'}(U_{k-1})$ the lines $\mathcal{L}_{n,R}^+$ and $\mathcal{L}_{n,R}^-$ intersect at $U_{k-1} < y_R^* < U_k$, here

$$y_R^* = \frac{(\alpha_{n'} - \alpha_n) - (\gamma_n + \gamma_n')R}{\beta_n - \beta_{n'}}. \tag{4.7}$$

n this case

$$\begin{aligned} m_k(n, n'; R) = &\lambda \Delta_{nn'}(\tilde{U}_k)(U_k - U_{k-1}) \\ &- \frac{\lambda}{2}(x_{n',R}^-(U_{k-1}) - x_{n,R}^+(U_{k-1}))(y^* - U_{k-1}), \end{aligned} \tag{4.8}$$

where $\tilde{U}_k = (U_{k-1}+U_k)/2$. Notice that $x^-_{n,R}(U_{k-1})-x^+_{n,R}(U_{k-1}) = \Delta_{nn'}(U_{k-1})-(\gamma_n+\gamma_{n'})R$. Thus, Eq. (4.8) can be written as

$$m_k(n,n';R) = \lambda\Delta_{nn'}(\tilde{U}_k)(U_k - U_{k-1}) - \frac{\lambda}{2}\frac{(\gamma_n+\gamma_{n'})^2}{\beta_n - \beta_{n'}}(R-\xi_{nn'}(U_{k-1}))^2. \tag{4.9}$$

Finally, if $R > \xi_{nn'}(U_{k-1})$ then

$$m_k(n,n';R) = \lambda\Delta_{nn'}(\tilde{U}_k)(U_k - U_{k-1}). \tag{4.10}$$

Thus, from Eqs. (4.6), (4.8) and (4.10) we obtain

$$\begin{aligned} m_k(n,n') = {} & \lambda(\gamma_n+\gamma_{n'})(U_k-U_{k-1})E\{RI\{R<\xi_{nn'}(U_k)\}\} \\ & + \lambda\Delta_{nn'}(\tilde{U}_k)(U_k-U_{k-1})P\{\xi_{nn'}(U_k) < R < \xi_{nn'}(U_{k-1})\} \\ & - \frac{\lambda}{2}\frac{(\gamma_n+\gamma_{n'})^2}{\beta_n-\beta_{n'}}E\{(R-\xi_{nn'}(U_{k-1}))^2 I\{\xi_{nn'}(U_k) < R < \xi_{nn'}(U_{k-1})\}\} \\ & + \lambda\Delta_{nn'}(\tilde{U}_k)(U_k-U_{k-1})P\{\xi_{nn'}(U_{k-1}) \le R\}. \end{aligned} \tag{4.11}$$

A similar formula for $m_k(n,n')$ can be obtained for the case of $\beta_{n'} > \beta_n$. In order to have a general formula for the two cases of $\beta_n \neq \beta_{n'}$, let $\xi^*_{nn'} = \min\{\xi_{nn'}(U_{k-1}), \xi_{nn'}(U_k)\}$ and $\xi^{*}_{nn'} = \max\{\xi_{nn'}(U_{k-1}), \xi_{nn'}(U_k)\}$. Then

$$\begin{aligned} m_k(n,n') = {} & \lambda(\gamma_n+\gamma_{n'}(U_k-U_{k-1})E\{RI\{R<\xi^*_{nn'}\}\} \\ & + \lambda\Delta_{nn'}(\tilde{U}_k)(U_k-U_{k-1})P\{\xi^*_{nn'} < R < \xi^{**}_{nn'}\} \\ & - \frac{\lambda}{2}\frac{(\gamma_n+\gamma_{n'})^2}{|\beta_n-\beta_{n'}|}E\{(R-\xi^{**}_{nn'})^2 I\{\xi^*_{nn'} < R < \xi^{**}_{nn'}\}\} \\ & + \lambda\Delta_{nn'}(\tilde{U}_k)(U_k-U_{k-1})P\{R \ge \xi^{**}_{nn'}\}. \end{aligned} \tag{4.12}$$

In the two equations above,

$$P\{\xi^*_{nn'} < R < \xi^{**}_{nn'}\} = F\{\xi^{**}_{nn'}) - F(\xi^*_{nn'}),$$

and

$$E\{R-\xi^{**}_{nn'})^2 I\{\xi^*_{nn'} < R < \xi^{**}_{nn'}\}\} = \int_{\xi^*_{nn'}}^{\xi^{**}_{nn'}} (r-\xi^{**}_{nn'})^2 f(r)dr. \tag{4.13}$$

In many of the examples of the previous chapters we considered visibility probabiliti[e]s when the radii, R, of random disks have uniform distributions, $\mathcal{U}(a,b)$. We provide no[w] the formulae for $m_k(n,n')$ when the distribution of R is uniform on $(0,b)$. First, in th[e] case of $\beta_n = \beta_{n'}$, Eq. (4.5) obtains the form:

$$m_k(n,n') = \begin{cases} \lambda\gamma_n(U_k-U_{k-1})b, & \text{if } d^* \ge 2b \\ \lambda\gamma_n(U_k-U_{k-1})d^*(1-\frac{d^*}{4b}), & \text{if } d^* < 2b \end{cases} \tag{4.14}$$

In the case of $\beta_n \neq \beta_{n'}$, Eq. (4.12) becomes

$$
\begin{aligned}
m_k(n,n') &= \lambda(\gamma_n + \gamma_{n'})(U_k - U_{k-1})\frac{b}{2}\left(\min\left(1, \frac{\xi^*_{nn'}}{b}\right)\right)^2 \\
&+ \lambda\Delta_{nn'}(\tilde{U}_k)(U_k - U_{k-1})\left(1 - \frac{\xi^*_{nn'}}{b}\right)^+ \\
&- \frac{\lambda}{2}\frac{(\gamma_n + \gamma_{n'})^2}{|\beta_n - \beta_{n'}|b}\left[\frac{1}{3}(\min(b, \xi^{**}_{nn'})^3 - \min(b, \xi^*_{nn'})^3)\right. \\
&- \xi^{**}_{nn'}(\min(b, \xi^{**}_{nn'})^2 - \min(b, \xi^*_{nn'})^2) \\
&+ \left.(\xi^{**}_{nn'})^2(\min(b, \xi^{**}_{nn'}) - \min(b, \xi^*))\right].
\end{aligned}
\tag{4.15}
$$

Example 4.1. We consider here $m = 3$ target points and $n = 3$ observation points. The scale of the coordinates is 1[m].

The coordinates of the target points and the observation points are

target points	(-20,100)	(0,100)	(50,100)
observation points	(-15,0)	(5,0)	(75,0)

The obscuring disks are dispersed in a strip between $u = 50$ and $w = 75$. The intensity of the field is $\lambda = 0.001[1/\text{m}^2]$ and the radii of random disks are uniformly distributed on the interval $(0, 1)$.

The program MOMTVPB computes the simultaneous visibility probability of the three target points from the three observation points. The computer program yields an output file MOMTVPB.DAT in which one can find the values α_n, β_n, γ_n of the N lines of sight; the y values at which these lines intersect inside the strip $\mathcal{S}$; and the order of the lines of sight within each substrip. In the present example we have $N = 9$ lines of sight. Their parameters are given in the following table.

Table 4.2. Parameters of Lines of Sight

n	α_n	β_n	γ_n
1	-15.0	-0.05	1.0012
2	-15.0	0.15	1.0112
3	-15.0	0.65	1.1927
4	5.0	-0.25	1.0308
5	5.0	-0.05	1.0012
6	5.0	0.45	1.0966
7	75.0	-0.95	1.3793
8	75.0	-0.75	1.2500
9	75.0	-0.25	1.0308

he y values at which these lines intersect within $\mathcal{S}$ are 56.25, 58.33 and 64.29. Ac-rdingly $\mathcal{S}$ is partitioned to $K = 4$ substrips. The order of the lines of sight within the bstrips, from left to right (clockwise) is:

Table 4.3. Order of Lines of Sight Within Substrips

k	ordered			line		indices			
1	1	4	2	5	3	7	6	8	9
2	1	4	2	5	7	3	6	8	9
3	1	4	2	5	7	3	8	6	9
4	1	4	2	5	7	8	3	6	9

The simultaneous visibility probability of all $N = 9$ lines is $\psi = 0.784$.

The program MOMTVPB can be used repeatedly to obtain simultaneous visibility probabilities of subsets of lines. For example, in the following table we present the simultaneous visibility probabilities of the three target points from each observation point individually, from pairs of observation points, and from all three.

Table 4.4. Joint Visibility Probabilities

Observation Points	Visibility Probability
1	0.923
2	0.925
3	0.913
1,2	0.855
1,3	0.845
2,3	0.846
1,2,3	0.784

Example 4.2. Consider $m = 4$ target points and $n = 4$ observation points. As in the previous example let the scale be 1[m]. The strip $\mathcal{S}$ is between $u = 50$ and $w = 75$ $\lambda = 0.001[1/\text{m}^2]$ and $b = 1$. The coordinates of the target points and the observation points are at

$$\text{Target} : (-25, 100), (-15, 95), (0, 97.5), (50, 110)$$
$$\text{Observation} : (-15, 0), (-10, -10), (15, 5), (25, 25).$$

Program MOMTVPB shows that there are twelve intersection points of the 16 lines of sight within $\mathcal{S}$. The y levels of these intersection points are:

$$50.9,\ 51.2,\ 51.9,\ 55.8,\ 57.3,\ 60.3,\ 65.0,$$
$$66.7,\ 69.0,\ 72.7,\ 73.8,\ 74.8.$$

Thus, $\mathcal{S}$ is partitioned to $k = 13$ substrips. The order of the lines within the substrips given in the following table. The joint visibility probability of all 16 lines is $\psi = 0.68$

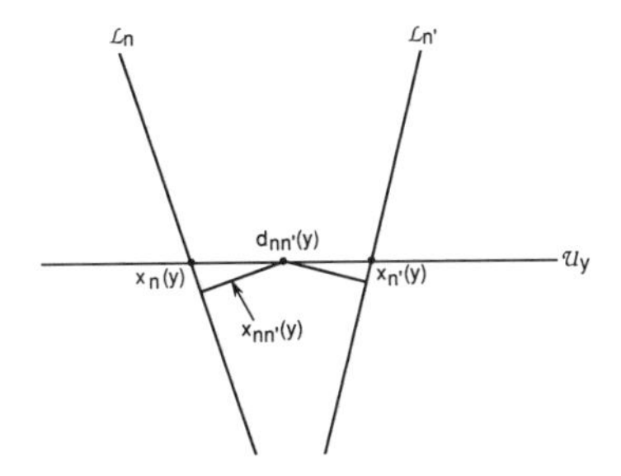

Figure 4.4. The Geometry of Intersecting Disks

Table 4.5. Order of Lines of Sight Within Substrips

k	Ordered Indexes of Lines of Sight															
1	1	5	2	6	3	7	9	10	11	13	14	4	15	8	12	16
2	1	5	2	6	3	9	7	10	11	13	14	4	15	8	12	16
3	1	5	2	6	3	9	7	10	13	11	14	4	15	8	12	16
4	1	5	2	6	3	9	7	10	13	11	14	15	4	8	12	16
5	1	5	2	6	9	3	7	10	13	11	14	15	4	8	12	16
6	1	5	2	6	9	3	7	10	13	14	11	15	4	8	12	16
7	1	5	2	6	9	3	10	7	13	14	11	15	4	8	12	16
8	1	5	2	6	9	10	3	7	13	14	11	15	4	8	12	16
9	1	5	2	6	9	10	3	13	7	14	11	15	4	8	12	16
10	1	5	2	6	9	10	13	3	7	14	11	15	4	8	12	16
11	1	5	2	6	9	10	13	3	14	7	11	15	4	8	12	16
12	1	5	2	9	6	10	13	3	14	7	11	15	4	8	12	16
13	1	5	2	9	6	10	13	14	3	7	11	15	4	8	12	16

■

4.3. Determination of $m_k(n,n')$ in Cases of Non-Standard Poisson Fields

For a given value of y in the interval $[U_{k-1}, U_k]$, let $x_n(y)$ $(n = 1, \cdots, N)$ denote the x-coordinate of the intersection points of the line $\mathcal{L}_n$ with the horizontal line at level y. Let $x_{nn'}(y)$ be the x-coordinate of the point on this horizontal line, having equal distance from $\mathcal{L}_n$ and $\mathcal{L}_{n'}$ (see Figure 4.4). It is clear that, for each $y \in (U_{k-1}, U_k)$,

$$x_n(y) < x_{nn'}(y) < x_{n'}(y). \tag{4.16}$$

For any point (x,y), let $d_n(x,y)$ denote its distance from $\mathcal{L}_n$ $(n = 1, \cdots, N)$. The formula of $m_k(n,n')$ is given, for $1 \leq n < n' \leq N$, by

$$
\begin{aligned}
m_k(n,n') = \lambda \int_{U_{k-1}}^{U_k} dy \Bigg[&\int_{x_n(y)}^{x_{nn'}(y)} h(x,y)(1 - F(d_n(x,y) \mid x,y))dx \\
+ &\int_{x_{nn'}(y)}^{x_{n'}(y)} h(x,y)(1 - F(d_{n'}(x,y) \mid x,y))dx \Bigg].
\end{aligned}
\tag{4.17}
$$

For $n = 0$ and $n' = 1, \cdots, N$, we have

$$
m_k(0,n') = \lambda \int_{U_{k-1}}^{U_k} dy \int_{-\infty}^{x_{n'}(y)} h(x,y)(1 - F(d_{n'}(x,y) \mid x,y))dx, \tag{4.18}
$$

and for $n = 1, \cdots, N$ we have

$$
m_k(n, N+1) = \lambda \int_{U_{k-1}}^{U_k} dy \int_{x_n(y)}^{\infty} h(x,y)(1 - F(d_n(x,y) \mid x,y))dx. \tag{4.19}
$$

In Eqs. (4.17)-(4.19), $h(x,y)$ is the p.d.f. of the disk centers in $\mathcal{S}$ and $F(r \mid x,y)$ is the conditional c.d.f. of the radius, R, given the location of the disk center. It is straightforward to verify that

$$
d_n(x,y) = |x - x_n(y)|/\gamma_n, \tag{4.20}
$$

and

$$
\begin{aligned}
x_{nn'}(y) &= (\gamma_n x_{n'}(y) + \gamma_{n'} x_n(y))/(\gamma_n + \gamma_{n'}) \\
&= \alpha_{nn'} + \beta_{nn'} y,
\end{aligned}
\tag{4.21}
$$

where

$$
\alpha_{nn'} = \frac{\gamma_{n'}\alpha_n + \gamma_n \alpha_{n'}}{\gamma_n + \gamma_{n'}},
$$

and

$$
\beta_{nn'} = \frac{\gamma_{n'}\beta_n + \gamma_n \beta_{n'}}{\gamma_n + \gamma_{n'}}.
$$

Let $d_{nn'}(y)$ be the distance of $x_{nn'}(y)$ from $\mathcal{L}_n$ (or from $\mathcal{L}_{n'}$). This distance is given b

$$
d_{nn'}(y) = \frac{x_{nn'}(y) - x_n(y)}{\gamma_n} = \frac{\Delta_{nn'}(y)}{\gamma_n + \gamma_{n'}}. \tag{4.22}
$$

In the notation of the previous section, $d_{nn'}(y) = \xi_{nn'}(y)$. Thus, if $F(r \mid x,y) = F(r$ Eq. (4.17) can be expressed in the form

$$
\begin{aligned}
m_k(n,n') = \lambda \int_{U_{k-1}}^{U_k} \Bigg[&\gamma_n \int_0^{\xi_{nn'}(y)} h(\gamma_n \rho + x_n(y), y)(1 - F(\rho))d\rho \\
+ \; &\gamma_{n'} \int_0^{\xi_{nn'}(y)} h(-\gamma_{n'}\rho + x_{n'}(y), y)(1 - F(\rho))d\rho \Bigg] dy.
\end{aligned}
\tag{4.23}
$$

Similarly, for $n = 1, \cdots, N$

$$m_k(0,n) = \lambda\gamma_n \int_{U_{k-1}}^{U_k} \int_0^{\infty} h(-\gamma_n\rho + x_n(y), y)(1 - F(\rho))d\rho dy \tag{4.24}$$

and

$$m_k(n, N+1) = \lambda\gamma_n \int_{U_{k-1}}^{U_k} \int_0^{\infty} h(\gamma_n\rho + x_n(y), y)(1 - F(\rho))d\rho dy. \tag{4.25}$$

More explicit formulae can be derived for special models. In particular, if we substitute $h(x,y) = 1$ and $F(\rho) = \min(1, \rho/b)$, we obtain the formulae of Section 4.2.

4.4. Joint Visibility of Windows

In the present section we generalize the results of Section 4.2 to the case where a whole interval of length L should be observable around each target point. Without loss of generality, assume that the window intervals are parallel to the x-axis (see Figure 4.5). From an observation point $\mathbf{O}_i$ we draw two lines to the boundaries of the interval around $\mathbf{T}_j$. Let $\mathcal{L}_{ij}^{(1)}$ and $\mathcal{L}_{ij}^{(2)}$ denote these two lines. Let us denote the parameters of $\mathcal{L}_{ij}^{(1)}$ and $L_{ij}^{(2)}$, respectively, by $\alpha_{ij}^{(1)}$, $\alpha_{ij}^{(2)}$ and $\beta_{ij}^{(1)}$, $\beta_{ij}^{(2)}$ ($i = 1, \cdots, m$, $j = 1, \cdots, n$), where $\beta_{ij}^{(1)} > \beta_{ij} > \beta_{ij}^{(2)}$. Notice that if (x_i^0, y_i^0), $i = 1, \cdots, n$ are the coordinates of the observation points, and (x_j^t, y_j^t), $j = 1, \cdots, m$, are those of the target points then

$$\beta_{ij}^{(1)} = \beta_{ij} + \frac{L}{2(y_j^t - y_j^0)},$$

and (4.26)

$$\beta_{ij}^{(2)} = \beta_{ij} - \frac{L}{2(y_j^t - y_i^0)}.$$

Similarly,

$$\alpha_{ij}^{(1),(2)} = x_j^t \pm \frac{L}{2} - \beta_{ij}^{(1),(2)} y_j^t. \tag{4.27}$$

As in Section 4.1, we order the lines $\mathcal{L}_{ij}$ as $\mathcal{L}_1, \cdots, \mathcal{L}_N$.

Consider the case of a standard Poisson field. Let $\psi(\mathcal{L}_1, \cdots, \mathcal{L}_N; L)$ denote the joint visibility probability with windows of length L. Eq. (4.1) is modified to

$$\psi(\mathcal{L}_1, \cdots, \mathcal{L}_N; L) = \exp\left\{-\sum_{k=1}^{K} M_k^{(N)}(L)\right\}, \tag{4.28}$$

here

$$M_k^{(N)}(L) = m_k(0,1;L) + \sum_{n=1}^{N-1} m_k(n, \nu_k^+(n); L) + m_k(N, N+1; L). \tag{4.29}$$

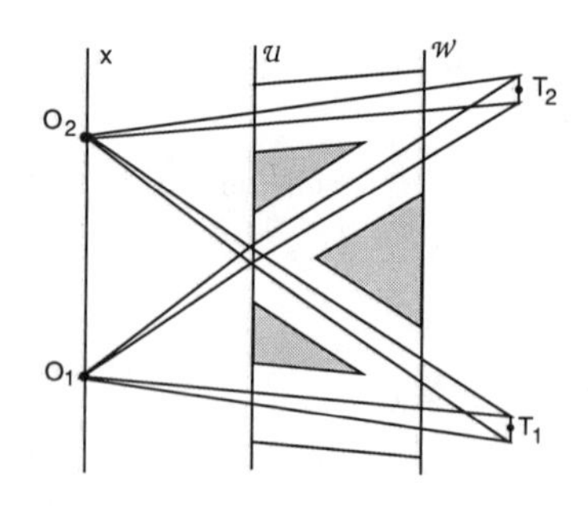

Figure 4.5. The Geometry of Target Points With Window Intervals

Let $m_k(0,n \mid \alpha_n, \beta_n)$, $m_k(n,n' \mid \alpha_n, \beta_n, \alpha_{n'}, \beta_{n'})$ and $m_k(n', N+1 \mid \alpha_{n'}, \beta_{n'})$ be the expressions for $m_k(0,n)$, $m_k(n,n')$ and $m_k(n', N+1)$, which are given in Section 4.2. As seen in Figure 4.5, the following modifications yield formulae when $L > 0$.

$$\begin{aligned} m_k(0,1;L) = m_k(0,1 \mid \alpha_1^{(2)}, \beta_1^{(2)}) + \lambda[(\alpha_1 - \alpha_1^{(2)})(U_k - U_{k-1}) \\ + \frac{1}{2}(\beta_1 - \beta_1^{(2)})(U_k^2 - U_{k-1}^2)], \end{aligned} \tag{4.30}$$

For $n' = \nu_k^+(n)$,

$$\begin{aligned} m_k(n,n';L) = m_k(n,n') \mid \alpha_n^{(1)}, \beta_n^{(1)}, \alpha_{n'}^{(2)}, \beta_{n'}^{(2)}) \\ + \lambda[(\alpha_n^{(1)} - \alpha_n + \alpha_{n'} - \alpha_{n'}^{(2)})(U_k - U_{k-1}) \\ + \frac{1}{2}(\beta_n^{(1)} - \beta_n + \beta_{n'} - \beta_{n'}^{(2)})(U_k^2 - U_{k-1}^2)], \end{aligned} \tag{4.31}$$

Finally,

$$\begin{aligned} m_k(N,N+1;L) = m_k(N,N+1) \mid \alpha_N^{(1)}, \beta_N^{(1)}) \\ + \lambda[(\alpha_N^{(1)} - \alpha_N)(U_k - U_{k-1}) + \frac{1}{2}(\beta_N^{(1)} - \beta_N)(U_k^2 - U_{k-1}^2)]. \end{aligned} \tag{4.32}$$

Example 4.3. Consider the set up of Example 4.1. The joint visibility of the $m =$ [illegible] target points from the $n = 3$ observation points, with windows of size $L = 0, 1, 2, 3$[m] can be computed by using program MOMTVPW. For a standard Poisson field centere[illegible] in the strip $\mathcal{S}$, between $u = 50$[m] and $w = 70$[m], with various intensity values λ, w[illegible] obtain the following joint visibility probabilities.

Table 4.6. Joint Visibility Probabilities

λ	$L = 0$	$L = 1$	$L = 2$	$L = 3$
0.001	0.784	0.686	0.601	0.530
0.005	0.297	0.152	0.079	0.042

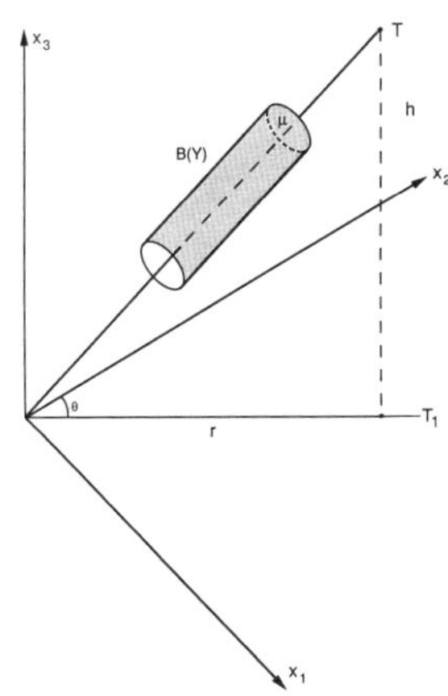

Figure 4.6. The Geometry of $B(Y)$ in Space

4.5. Visibility of Points in Space

In the present section we develop formulae for some visibility probabilities of points in a three dimensional space. We assume that the observer is located at the origin **O**, and the obscuring elements are random spheres, centered between two horizontal planes $\mathcal{U}$ and $\mathcal{W}$, of distance u and w from the origin $(0 < u < w < \infty)$. Furthermore, we assume that the random spheres constitute a standard Poisson field. A sphere radius, Y, has a specified distribution $F(y)$, on a finite interval (a, b), where $0 < a < b < u$. We start with the visibility probability of a single target.

4.5.1. Single Target

We specify the target point **T** by its cylindrical coordinates (r, θ, h), where (r, θ) are the polar coordinates of the (right) projection. $\mathbf{T}'$ of **T** on the plane $\mathbb{C}$ spanned by the x_1-, x_2-axes. The vertical height of **T** (distance of **T** from $\mathcal{C}$) is h, $h > 0$. We assume that $h > w$ (see Figure 4.6).

The cylinder $B(Y)$, is the set of all centers of spheres of radius Y that could intersect the line of sight $\overline{\mathbf{OT}}$. It is straightforward to show that the volume of $B(Y)$ is

$$\#\{B(Y)\} = (w-u)\left(1+\left(\frac{r}{h}\right)^2\right)^{1/2}\pi Y^2. \tag{4.33}$$

Accordingly, if μ_2 denotes the second moment of the distribution of Y, and λ is the intensity of the Poisson field [1/m^3], then the visibility probability of the point **T** is

$$\psi(\mathbf{T}) = \exp\left\{-\lambda(w-u)\left(1+\left(\frac{r}{h}\right)^2\right)^{1/2}\pi\mu_2\right\}. \tag{4.34}$$

We see in Eq. (4.34) that if the centers of the random spheres are scattered in an unbounded layer in space, between the horizontal planes $\mathcal{U}$, $\mathcal{W}$, the visibility probability $\psi(T)$ is independent of the orientation coordinate θ, of **T**. Formula (4.34) can be extended to provide the visibility probability of a sphere of radius l, S_T, centered at **T** (three dimensional window).

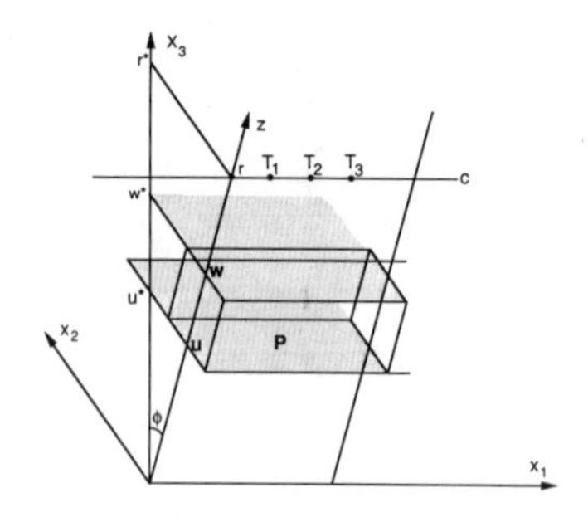

Figure 4.7. The Geometry of the Three Dimensional Model

Example 4.4. Consider an observation point $\mathbf{O}$ at the origin, and a target point at $\mathbf{T} = (200, \theta, 1000)$. Distances are measured in units of 1[m]. Suppose that a standard Poisson field of random spheres is located between the parallel planes $\mathcal{U}$, $\mathcal{W}$, where $u = 300$[m] and $w = 600$[m]. The radius of a sphere, Y, is uniformly distributed on $(0, 1)$[m]. In this case $\mu_1 = 0.5$[m], $\mu_2 = \frac{1}{3}$[m^2]. The intensity of the field is $\lambda = 0.001[1/\text{m}^3]$. Eq. (4.34) yields that the visibility probability of $\mathbf{T}$ is $\psi(\mathbf{T}) = 0.726$. ■

Visibility of windows around the target point will be discussed at the end of section 4.5.4.

4.5.2. Several Target Points On A Line

We consider now the problem of determining the simultaneous visibility probability of several target points on a straight line $\mathcal{C}$ in space (see Figure 4.7). The observation point is at the origin $\mathbf{O}$. The obscuring elements are random spheres centered in a layer, bounded between two parallel planes $\mathcal{U}^*$ and $\mathcal{W}^*$, which are parallel to $\mathcal{C}$. These spheres constitute a standard Poisson field in space. We denote by $F(y)$ the c.d.f. of the random diameter, Y, of a sphere. We assume that $a \leq Y \leq b < \infty$. Moreover, nc random sphere can cover the origin $\mathbf{O}$, or intersect the target line $\mathcal{C}$. Let u^*, w^* and r^* be the distance from $\mathbf{O}$ of the planes, $\mathcal{U}^*$, $\mathcal{W}^*$ and $\mathcal{C}^*$, where $\mathcal{C}^*$ is a plane parallel tc $\mathcal{W}^*$ to which $\mathcal{C}$ belongs. Let $\mathcal{M}$ be a plane passing through $\mathbf{O}$ and $\mathcal{C}$ (see Figure 4.7) Let $\mathbf{O}'$ be the right projection of $\mathbf{O}$ on $\mathcal{C}$ and let $\mathcal{Z}$ be the line connecting $\mathbf{O}$ and $\mathbf{O}'$ The inclination of the plane $\mathcal{M}$ is given by an angle ϕ between the x_3-axis ($\mathcal{Z}^*$) and $\mathcal{Z}$ Let $\mathcal{U}$ and $\mathcal{W}$ denote the straight lines at which $\mathcal{M}$ intersects the planes $\mathcal{U}^*$ and $\mathcal{W}^*$ respectively. The distances of $\mathcal{U}$, $\mathcal{W}$ and $\mathcal{C}$ from $\mathbf{O}$ are u, w and r, respectively. Sphere centered in the layer between $\mathcal{U}^*$ and $\mathcal{W}^*$ may intersect $\mathcal{M}$, only if they are centere at the prism $\mathcal{P}$ (see Figure 3.16), of width $2b$, around $\mathcal{M}$. The intersection of a spher with $\mathcal{M}$ is a circle. If such a circle intersects the line of sight between $\mathbf{O}$ and a point T on $\mathcal{C}$ then the point is not visible from $\mathbf{O}$. Accordingly, the three dimensional visibilit problem is reduced to a two dimensional visibility problem on $\mathcal{M}$. We develop belo the K-functions for the Poisson random field of obscuring circles on $\mathcal{M}$.

In Figure 4.8 we illustrate a cut of the prism $\mathcal{P}$ by a plane containing $\mathcal{Z}$ and $\mathcal{Z}^*$. Le $\mathcal{P}^*$ be the parallelogram $ABCD$, which is the cut of $\mathcal{P}$. The right projection of point in $\mathcal{P}^*$ on $\mathcal{Z}$ is the interval $(u - \beta, w + \beta)$, where $\beta = b \ \tan(\phi)$.

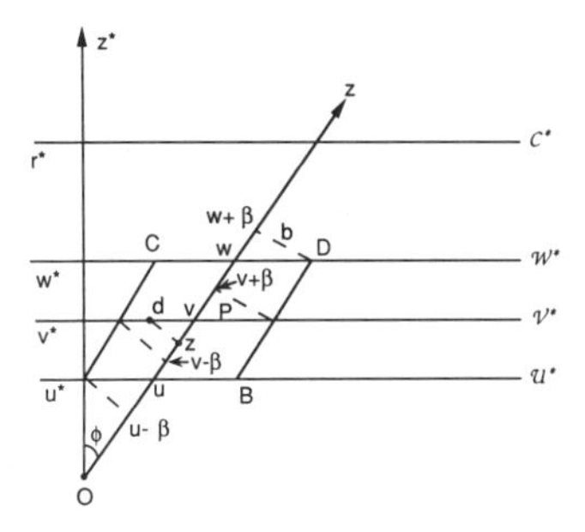

Figure 4.8. The Geometry of The Reduction to a Two-Dimensional Model

Consider a sphere centered in $\mathcal{P}$, on a plane $\mathcal{V}^*$, of distance v^* from $\mathbf{O}$ (see Figure 4.9). Let d be the distance of the center of this sphere from $\mathcal{M}$. If the radius, Y, of this disk is greater than d, then it intersects $\mathcal{M}$ with a disk, whose center is on a line of distance z from $\mathbf{O}$. Moreover

$$|z - v| = d\frac{\beta}{b}. \tag{4.37}$$

The radius of the intersecting disk on $\mathcal{M}$ is

$$\tilde{Y} = (Y^2 - (z - v)^2 b^2/\beta^2)_+^{1/2}, \tag{4.38}$$

where, generally $x_+ = \max(0, x)$. Accordingly, the conditional c.d.f. of G of $\tilde{Y}$ is

$$G(y \mid v, z) = \begin{cases} 0, & \text{if } y < 0 \\ F\left(\left(y^2 + b^2\left(\frac{z-v}{\beta}\right)^2\right)^{1/2}\right), & \text{if } 0 \le y \le b\left(1 - \left(\frac{z-v}{\beta}\right)^2\right)^{1/2} \\ 1, & \text{if } y > b\left(1 - \left(\frac{z-v}{\beta}\right)^2\right)^{1/2}. \end{cases} \tag{4.39}$$

Consider now n points on $\mathcal{C}$. Let $-\infty < \xi_1 < \xi_2 < \cdots < \xi_n < \infty$ be the cartesian coordinates of the target points $\mathbf{T}_1, \cdots, \mathbf{T}_n$, with respect to the point $\mathbf{O}'$ on $\mathcal{C}$. Let $\mathbf{P}_L$ and $\mathbf{P}_U$ be two points on $\mathcal{C}$, with coordinates $\xi' < \xi_1$ and $\xi'' > \xi_n$, respectively, such hat no disk on $\mathcal{M}$, centered to the left of $\overline{\mathbf{OP}}_L$, or to the right of $\overline{\mathbf{OP}}_U$, can obscure he visibilities of the target points. $\mathbf{P}_L$ and $\mathbf{P}_U$ can be chosen so that

$$\xi' = \xi_1 - b(\xi_1^2 + r^2)^{1/2}/(u - \beta)$$

nd

$$\xi'' = \xi_n + b(\xi_n^2 + r^2)^{1/2}/(u - \beta).$$

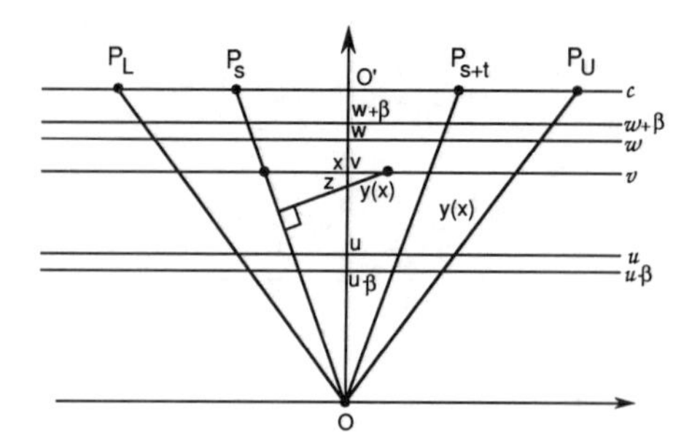

Figure 4.9. The Geometry of Lines of Sight on $\mathcal{M}$

Let C be the set of points on $\mathcal{M}$ bounded by the lines $\overline{\mathbf{OP}}_L$, $\overline{\mathbf{OP}}_U$, $\mathcal{U}-\beta$ and $\mathcal{W}+\beta$ (see Figure 4.9). The expected number of disks centered in C is $\lambda A\{C\}$, where

$$\begin{aligned} A\{C\} &= \frac{b}{\beta}\frac{\xi''-\xi'}{r}\int_u^w dv \int_{v-z}^{v+z} z dz \\ &= \frac{b}{r}(\xi''-\xi')(w^2-u^2). \end{aligned} \tag{4.40}$$

For a value of s in (ξ', ξ'') and $t < \xi''-\xi'-s$, $K_+(s,t)$ is the probability that a random disk on $\mathcal{M}$, centered between $\overline{\mathbf{OP}}_s$ and $\overline{\mathbf{OP}}_{s+t}$, does not intersect $\overline{\mathbf{OP}}_s$. Similarly, $K_-(s,t)$ is the probability that a disk on $\mathcal{M}$, centered between $\overline{\mathbf{OP}}_{s-t}$ and $\overline{\mathbf{OP}}_s$, does not intersect $\overline{\mathbf{OP}}_s$.

A disk centered between $\overline{\mathbf{OP}}_s$ and $\overline{\mathbf{OP}}_{s+t}$, on a line $\mathcal{V}$ parallel to $\mathcal{U}$, of distance z from $\mathbf{O}$, and whose center is at distance x from $\overline{\mathbf{OP}}_s$, will not intersect $\overline{\mathbf{OP}}_s$ if its radius is smaller than

$$y(x) = \frac{xr}{(s^2+r^2)^{1/2}}, \qquad 0 \le x \le \frac{z}{r}t. \tag{4.41}$$

Accordingly,

$$K_+(s,t) = \frac{\lambda b}{\mu\{C\}\beta}\int_u^w dv \int_{v-\beta}^{v+\beta} dz \int_0^{zt/r} G\left(\frac{xr}{(s^2+r^2)^{1/2}} \mid v, z\right) dx. \tag{4.42}$$

Geometrical considerations imply that $K_+(s,t) = K_-(s,t)$. We therefore denote these functions by $K(s,t)$. We will prove in the next section that

$$\begin{aligned} \frac{1}{\lambda}\mu\{C\}K(s,t) = \frac{b\beta^2 t}{c^2 r}\Bigg[&\frac{c^2}{\beta^2}(w^2-u^2) - \frac{\pi c}{\beta}(w-u)\Psi(0) \\ &+ Q\left(\frac{c}{\beta}u, c\right) - Q\left(\frac{c}{\beta}w, c\right)\Bigg], \end{aligned} \tag{4.43}$$

where $c = \alpha\beta t/b$, $\alpha = (s^2+r^2)^{-1/2}$,

$$\Psi(\tau) = \int_\tau^1 [1-F(b\rho)]\rho d\rho, \tag{4.44}$$

and

$$
Q(\nu, c) = \begin{cases} 0, & \text{if } \nu \geq (1+c^2)^{1/2} \\ 2(1+c^2)^{1/2} \int_{v/(1+c^2)^{1/2}}^{1} \frac{\eta d\eta}{(1-\eta^2)^{1/2}} \int_{v/\eta(1+c^2)^{1/2}}^{1} \Psi(z)dz, & \text{if } c \leq v \leq (1+c^2)^{1/2} \\ (1+c^2)^{1/2} \int_{v/(1+c^2)^{1/2}}^{1} \frac{\eta d\eta}{(1-\eta^2)^{1/2}} \left[\int_{v/\eta(1+c^2)^{1/2}}^{1} \Psi(z)dz \right. & \\ \left. + \int_{M(1,c/\eta(1+c)^{1/2})}^{1} \Psi(z)dz \right], & \text{if } v < c \end{cases} \tag{4.45}
$$

where $M(x,y) = \min\{x,y\}$. Finally, the simultaneous visibility probability of the points $\mathbf{T}_1, \cdots, \mathbf{T}_n$ on $\mathcal{C}$ is

$$
\begin{aligned} \psi(\xi_1, \cdots, \xi_n) = \exp\{-\lambda A\{C\}[1 - (K(\xi_1, \xi_1 - \xi') + K(\xi_n, \xi'' - \xi_n)) \\ - \sum_{i=1}^{n-1} (K(\xi_i, \tilde{\xi}_i - \xi_i) + K(\xi_{i+1}, \xi_{i+1} - \tilde{\xi}_i))]\} \end{aligned} \tag{4.46}
$$

where

$$
\tilde{\xi}_i = r \tan\left(\frac{1}{2}\left(\tan^{-1}\left(\frac{\xi_i}{r}\right) + \tan^{-1}\left(\frac{\xi_{i+1}}{r}\right)\right)\right). \tag{4.47}
$$

4.5.3. Uniform Distribution of Sphere Radius

In the present section we develop further the formulae of the previous section, for the case where the distribution of Y is uniform on $(0,b)$. In this case Eq. (4.44) gets the form

$$
\Psi(\tau) = \frac{1}{6} - \frac{\tau^2}{2} + \frac{\tau^3}{3}, \qquad 0 \leq \tau \leq 1. \tag{4.48}
$$

Let

$$
H(x,z) = \int_x^1 \frac{\eta}{(1-\eta^2)^{1/2}} \left(\int_{1/\eta}^1 \Psi(\tau)d\tau \right) d\eta \tag{4.49}
$$

From Eq. (4.48) we obtain

$$
\begin{aligned} H(x,z) &= \int_x^1 \frac{\eta}{(1-\eta^2)^{1/2}} \left(\int_{z/\eta}^1 \left(\frac{1}{6} - \frac{1}{2}\tau^2 + \frac{1}{3}\tau^3 \right) d\tau \right) d\eta \\ &= \frac{1}{12}\left[(1-x^2) - 2\pi + 2z\sin^{-1}(x) + 2z^3(1-x^2)^{1/2}/x - z^4 \left(\frac{(1-x^2)^{1/2}}{2x^2} \right.\right. \\ &\quad \left.\left. + \frac{1}{2}\log\frac{1+(1-x^2)^{1/2}}{x} \right)\right]. \end{aligned} \tag{4.50}
$$

Furthermore, from Eq. (4.45),

$$Q(v,c) = \begin{cases} 0, & \text{if } v \geq (1+c^2)^{1/2} \\ 2(1+c^2)^{1/2} H\left(\frac{v}{(1+c^2)^{1/2}}, \frac{v}{(1+c^2)^{1/2}}\right), & c \leq v \leq (1+c^2)^{1/2} \\ (1+c^2)^{1/2}\left[H\left(\frac{v}{(1+c^2)^{1/2}}, \frac{v}{(1+c^2)^{1/2}}\right)\right. & \\ \left. -H\left(\frac{c}{(1+c^2)^{1/2}}, \frac{v}{(1+c^2)^{1/2}}\right)\right], & \text{if } 0 \leq v < c \end{cases} \tag{4.51}$$

4.5.4. Derivation of $K(s,t)$

In the present section we derive formula (4.43). According to Eqs. (4.39) and (4.42),

$$K(s,t) = \frac{\lambda b}{\mu\{C\}\beta\alpha r}\int_u^w dv \int_{v-\beta}^{v+\beta} dz \int_0^{\alpha z t} F\left(\left(y^2 + \left(\frac{z-v}{\beta}\right)^2 b^2\right)^{1/2}\right) dy, \tag{4.52}$$

in which $\alpha = (s^2+r^2)^{-1/2}$.

Let $c = \alpha\beta t/b$; then, after several simple changes of variables, we can write

$$\begin{aligned} \frac{1}{\lambda}\mu\{C\}K(s,t) &= \frac{b\beta^2 t}{c^2 r}\int_{(c/\beta)u}^{(c/\beta)w} dy \int_{-1}^{1} dx \int_0^{y+cx} F(b(z^2+x^2)^{1/2})dz \\ &= \frac{b\beta^2 t}{c^2 r}\left[\frac{c^2}{\beta^2}(w^2-u^2) - J(c,\beta,u,w)\right], \end{aligned} \tag{4.53}$$

where

$$J(c,\beta,u,w) = \int_{(c/\beta)u}^{(c/\beta)w} dy \int_{-1}^{1} dx \int_0^{y+cx} [1 - F(b(x^2+z^2)^{1/2})]dz. \tag{4.54}$$

Notice that $F(b(x^2+z^2)^{1/2} = 1$ for all (x,z), s.t. $x^2+z^2 \geq 1$. Write

$$J(c,\beta,u,w) = J_1(c,\beta,u,w) - J_2(c,\beta,u,w), \tag{4.55}$$

where

$$J_1(c,\beta,u,w) = \int_{(c/\beta)u}^{(c/\beta)w} dy \int_{-1}^{1} dx \int_0^{(1-x^2)^{1/2}} [1 - F(b(x^2+z^2)^{1/2})]dz \tag{4.56}$$

and

$$J_2(c,\beta,u,w) = \int_{(c/\beta)u}^{(c/\beta)w} dy \int_{-1}^{1} dx \int_{M[y+cx,(1-x^2)^{1/2}]}^{(1-x^2)^{1/2}} [1 - F(b(x^2+z^2)^{1/2})]dx, \tag{4.57}$$

where $M(a,b) = \min(a,b)$.

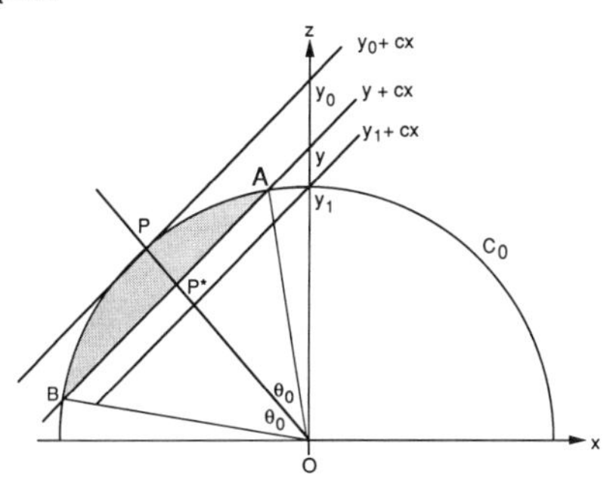

Figure 4.10. Integrated set, Case I.

Transforming (x, z) to be the polar coordinates (ρ, θ) we obtain

$$\begin{aligned} J_1(c, \beta, u, w) &= \int_{(c/\beta)u}^{(c/\beta)u} dy \int_0^{\pi} d\theta \int_0^1 \rho[1 - F(b\rho)]d\rho \\ &= \pi \frac{c}{\beta}(w - u)\Psi(0)). \end{aligned} \tag{4.58}$$

where $\Psi(t)$ is defined in (4.44).

In order to evaluate the function $J_2(c, \beta, u, w)$, define the function

$$Q(v, c) = \int_v^x dy \int_{-1}^1 dx \int_{y+cx}^x [1 - F(b(x^2 + z^2)^{1/2})]dz. \tag{4.59}$$

Then,

$$J_2(c, \beta, u, w) = Q\left(\frac{c}{\beta}u, c\right) - Q\left(\frac{c}{\beta}w, c\right). \tag{4.60}$$

Consider the half circle $C_0 = \{(x, z) : -1 \leq x \leq 1,\ 0 \leq z \leq 1,\ x^2 + z^2 \leq 1\}$ (see 'igure 4.10).

Let $\mathcal{L}$ be the line $z = y + cx$, having a slope c, $c > 0$, and intercept y. Let $\mathcal{L}_0$ be a line arallel to $\mathcal{L}$, which is tangential to C_0. Let y_0 be the intercept of $\mathcal{L}_0$. Similarly, let $\mathcal{L}_1$ e a line parallel to $\mathcal{L}$, passing through the point $(-1, 0)$. $\mathcal{L}_1$ has an intercept $y_1 = c$. et **P** be the point at the intersection of C_0 and $\mathcal{L}_0$. The right triangle with vertices **O**, **P**, $(0, y_0)$ is congruent to the triangle $(-1, 0)$, **O**, $(0, y_1)$. Hence $y_0 = (1 + c^2)^{1/2}$. It clear that if $v > y_0$, then $Q(v, c) = 0$. We distinguish two cases.

Case I ($y_1 \leq v < y_0$): The line segment $\overline{\mathbf{OP}}$ intersects the line $\mathcal{L}$ at $\mathbf{P}^*$, whose stance from **O** is $\gamma = y/y_0$ (see Figure 4.11). Let **A** and **B** be the points at which $\mathcal{L}$ tersects the half circle C_0. The triangle ΔBOA is equilateral and

$$\theta_0 = \not\measuredangle\, \mathbf{POA} = \cos^{-1}(\gamma). \tag{4.61}$$

ence, by changing to polar coordinates and making a proper rotation, we obtain

$$Q(v, c) = 2 \int_v^{y_0} dy \int_0^{\cos^{-1}(y/y_0)} \Psi\left(\frac{y}{y_0 \cos\theta}\right) d\theta. \tag{4.62}$$

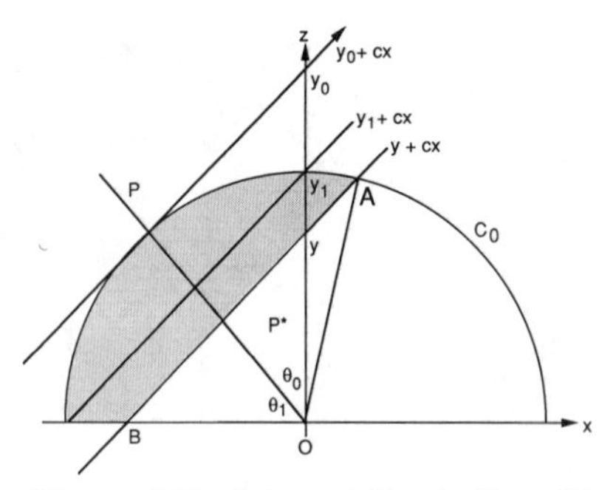

Figure 4.11. Integrated set, Case II.

Case II ($0 \leq v < y_1$): Let θ_1 denote the angle between $\overline{\mathbf{OP}}$ and $\overline{\mathbf{OB}}$ (see Figure 4.11). It is immediately obtained that $\theta_1 = (\pi/2) - \tan^{-1}(c) = \sin^{-1}(1/\sqrt{1+c^2}) = \tan^{-1}(1/c)$. Thus, in Case II,

$$Q(v,c) = \int_v^{y_0} dy \left[\int_0^{\theta_0} \Psi\left(\frac{y}{y_0 \cos\theta}\right) d\theta + \int_0^{\theta_1} \Psi\left(\frac{y}{y_0 \cos\theta}\right) d\theta \right]. \tag{4.63}$$

Making the transformation $\eta = \cos(\theta)$, $z = y/y_0\eta$, and changing the order of integration, we obtain

$$Q(v,c) = \begin{cases} 0, & v \geq y_0 \\ 2_{y_0} \displaystyle\int_{v/y_0}^1 \frac{\eta d\eta}{\sqrt{1-\eta^2}} \int_{v/y_0\eta}^1 \Psi(z)dz, & y_1 \leq v < y_0 \\ y_0 \displaystyle\int_{v/y_0}^1 \frac{\eta d\eta}{\sqrt{1-\eta^2}} \left[\int_{v/y_0\eta}^1 \Psi(z)dz + \int_{M(y_1/y_0\eta,1}^1 \Psi(z)dz \right], & 0 \leq v < y_1, \end{cases} \tag{4.64}$$

Notice that in the case of $0 < v < y_1$, one can write

$$\begin{aligned} Q(v,c) = y_0 \Bigg[& \int_{v/y_0}^1 \frac{\eta d\eta}{\sqrt{1-\eta^2}} \int_{v/y_0\eta}^1 \Psi(z)dz \\ & + \int_{y_1/y_0}^1 \frac{\eta d\eta}{\sqrt{1-\eta^2}} \int_{y_1/y_0\eta}^1 \Psi(z)dz \Bigg]. \end{aligned} \tag{4.65}$$

Finally, according to (4.53), (4.55), (4.58) and (4.60)

$$\begin{aligned} \frac{1}{\lambda}\mu\{C\}K(s,t) = \frac{b\beta^2 t}{c^2 r} \Bigg[& \frac{c^2}{\beta^2}(w^2 - u^2) - \frac{\pi c}{\beta}(w-u)\Psi(0) \\ & + Q\left(\frac{c}{\beta}u, c\right) - Q\left(\frac{c}{\beta}w, c\right) \Bigg]. \end{aligned} \tag{4.66}$$

If intervals of length L (windows) should be completely observable around the targ points we can obtain the joint visibility probabilities by using the previous formul the modification is in the s and t values of the functions $K(s,t)$. Eq. (4.46) has to be changed to

$$\psi_w(\xi_1,\cdots,\xi_n;L) = \exp\{-\lambda A\{C\}\cdot \Bigg[1-\left(K\left(\xi_1-\frac{L}{2},\left(\xi_1-\xi_1'-\frac{L}{2}\right)^+\right)+K\left(\xi_n+\frac{L}{2},\left(\xi''-\xi_n-\frac{L}{2}\right)^+\right)\right) - \sum_{i=1}^{n-1}\left(k\left(\xi_i+\frac{L}{2},\left(\tilde{\xi}_i-\xi_i-\frac{L}{2}\right)^+\right)+K\left(\xi_{i+1}-\frac{L}{2},\left(\xi_{i-1}-\tilde{\xi}_i-\frac{L}{2}\right)^+\right)\right)\Bigg] \tag{4.67}$$

where $(a)^+ - \max(a,0)$ and $K(s,0) \equiv 0$. The computation of the visibility probability for n points on the line C, can be performed with the aid of the program THRDVPW as illustrated in the following example.

Example 4.5. Consider $n = 4$ target points in space. The ξ-coordinates (in [m]) of these points on the target line C are: -15, -7, 0, 8. The elevation parameters are $u^* = 50$[m], $w^* = 75$[m] and $r^* = 100$[m]. The end points of interval C are $\xi_L = -25$[m] and $\xi_U = 25$[m]. The intensity of the Poisson field is $\lambda = 0.001[1/\text{m}^3]$. The radii of random spheres are uniformly distributed in the interval $(0, b)$. The inclination angle is ϕ. In the following table we present the simultaneous visibility probabilities as functions of b, ϕ and the window size L. These values are computed with program THRDVPW.

Table 4.7. 3D-Visibility Probabilities

L[m]	b[m]	$\phi = 15°$	$\phi = 45°$	$\phi = 60°$
0	1	0.8968	0.8621	0.8109
0	3	0.4226	0.3132	0.1967
1	1	0.8009	0.7385	0.6515
1	3	0.2208	0.1275	0.0545

■

5

Distributions of Visibility Measures

In the present chapter we develop methods for determining the distributions of some measures of visibility for a given field of obstacles. We start first with the distribution of the number of targets which are simultaneously visible, out of m specified target points, either from one or from several observation points. This distribution is *not* the frequently encountered binomial distribution, since the visibility for specified targets are generally not independent, and the visibility probabilities are generally not the same. The other measure of visibility studied in this chapter is an integrated measure of the total length, of a specified star-shaped curve, that can be observed from one or several observation points. These measures have various applications.

5.1. The Distribution of the Number of Visible Targets

5.1.1. Introductory Examples With One Observation Point

Let $\{\mathbf{T}_i;\ i = 1, \cdots, m\}$ be an ordered set of m specified target points.

Let $\{O_i\}$ denote the event that the i-th point is visible (observable) and $\{O_i^{-1}\}$ the (complementary) event that $\mathbf{T}_i$ is invisible. Let $\psi_1\{O_i\}$ be the (marginal) probability of this event. Obviously, $\psi_1\{O_i^{-1}\} = 1 - \psi_1\{O_i\}$. Similarly, let $\{O_{i_1} O_{i_2}\}$ be the event that two points $\mathbf{T}_{i_1}$ and $\mathbf{T}_{i_2}$ ($i_1 \neq i_2$) are simultaneously visible. Let $\{O_{i_1} O_{i_2}^{-1}\}$ be the event that $\mathbf{T}_{i_1}$ is visible **and** $\mathbf{T}_{i_2}$ is invisible; $\{O_{i_1}^{-1} O_{i_2}\}$ is the event that $\mathbf{T}_{i_1}$ is invisible and $\mathbf{T}_{i_2}$ is visible, and so on. We wish to determine the probabilities of all the 2^m elementary events $\{O_1^{j_1} O_2^{j_2} \cdots O_m^{j_m}\}$, where $j_1, j_2, \cdots, j_m = \pm 1$. The methods developed in Chapters 3 and 4 yield the joint **visibility** probabilities of any subset of the m target points. By applying the basic rules of probability theory, we can determine from the m joint visibility probabilities of all the $2^m - 1$ subsets of $\{T_1, \cdots, T_m\}$ (excluding the empty set), the probabilities of above elementary events.

Example 5.1. Let $\mathbf{T}_1$, $\mathbf{T}_2$, $\mathbf{T}_3$ be three specified target points. Let $\psi_1(i)$, $i = 1, 2, 3$ be their visibility probabilities; let $\psi_2(i_1, i_2)$, $i \leq i_1 < i_2 \leq 3$, be the simultaneous visibility probabilities of the three pairs of points, and let $\psi_3(1, 2, 3)$ be the simultaneous visibility probability of all the three points.

Let $\Pr\{O_1^{j_1} O_1^{j_2}\}$ be the probability of the four possible events among two points when $j_i = \pm 1$. Then,

$$\Pr\{O_{i_1} O_{i_2}\} = \psi_2(i_1, i_2)$$

$$\Pr\{O_{i_1} O_{i_2}^{-1}\} = \psi_1(i_1) - \psi_2(i_1, i_2)$$

$$\Pr\{O_i^{-1} O_i\} = \psi_1(i_2) - \psi_2(i_1, i_2)$$

and

$$\Pr\{O_i^{-1} O_i^{-1}\} = 1 - \psi_1(i_1) - \psi_1(i_2) + \psi_2(i_1, i_2).$$

Now, when we consider all the eight elementary events, we get

$$\begin{aligned}\Pr\{O_1O_2O_3\} &= \psi_3(1,2,3),\\ \Pr\{O_1^{-1}O_2O_3\} &= \Pr\{O_2O_3\} - \Pr\{O_1O_2O_3\}\\ &= \psi_2(2,3) - \psi_3(1,2,3).\end{aligned}$$

Similarly,

$$\begin{aligned}\Pr\{O_1O_2^{-1}O_3\} &= \psi_2(1,3) - \psi_3(1,2,3)\\ \Pr\{O_1^{-1}O_2^{-1}O_3) &= \Pr\{O_2^{-1}O_3\} - \Pr\{O_1O_2^{-1}O_3\}\\ &= \psi_1(3) - \psi_2(2,3) - \psi_2(1,3) + \psi_3(1,2,3)\\ \Pr\{O_1O_2O_3^{-1}\} &= \psi_2(2,2) - \psi_3(1,2,3)\\ \Pr\{O_1^{-1}O_2O_3^{-1}\} &= \Pr\{O_2O_3^{-1}\} - \Pr\{O_1O_2O_3^{-1}\}\\ &= \psi_1(2) - \psi_2(2,3) - \psi_2(1,2) + \psi_3(1,2,3)\\ \Pr\{O_1O_2^{-1}O_3^{-1}\} &= \Pr\{O_1O_2^{-1}\} - \Pr\{O_1O_2^{-1}O_3\}\\ &= \psi_1(1) - \psi_2(1,2) - \psi_2(1,3) + \psi_3(1,2,3)\end{aligned}$$

nd finally,

$$\begin{aligned}\Pr\{O_1^{-1}O_2^{-1}O_3^{-1}\} = 1 &- \psi_1(1) - \psi_1(2) - \psi_1(3) + \psi_2(1,2)\\ &+ \psi_2(1,3) + \psi_2(2,3) - \psi_3(1,2,3).\end{aligned}$$

■

A general algorithm for obtaining the probability of an elementary event $O_1^{j_1}O_2^{j_2}\cdots O_m^{j_m}$ an be obtained in the following manner:

Step 1. Expand the product $\prod_{i=1}^{m}(I(j_i) + (-1)^{I(j_i)}O_i)$ where $I(j) = 0$ if $j = 1$ and $I(j) = 1$ if $j = -1$.

Step 2. Replace every term of the form $O_{i_1}O_{i_2}\cdots O_{i_k}$ in this expansion by the visibility probabilities $\psi_k\ (i_1, i_2, \cdots, i_k)$.

xample 5.2. Consider the elementary event $O_1O_2^{-1}O_3O_4^{-1}$. In Step 1 we expand

$$\begin{aligned}O_1(1-O_2)O_3(1-O_4) &= O_1O_3(1 - O_2 - O_4 + O_2O_4)\\ &= O_1O_3 - O_1O_2O_3 - O_1O_3O_4 + O_1O_2O_3O_4.\end{aligned}$$

cordingly,

$$\begin{aligned}\Pr\{O_1O_2^{-1}O_3O_4^{-1}\} = \psi_2(1,3) &- \psi_3(1,2,3) - \psi_3(1,3,4)\\ &+ \psi_4(1,2,3,4).\end{aligned}$$

■

After determining the probabilities of all the 2^m elementary events one can proceed to determine the distribution of the number of visible targets J. For this purpose, classify all the elementary events into $m+1$ subsets

$$A_l = \{O_1^{j_1} O_2^{j_2} \cdots O_m^{j_m};\ j_i = \pm 1\ (i = 1, \cdots, m) \text{ and } \sum_{i=1}^{m} j_1 = 2l - m\},\quad l = 0, 1, \cdots, m.$$

Then, add the elementary probabilities $\Pr\{O_1^{j_1} O_2^{j_2} \cdots O_m^{j_m}\}$ of all the elements of A_l to obtain $P_r\{J = l\}$.

Example 5.3. In the case of $m = 3$ targets, and $n = 1$ observation point we have 4 possible sets:

$$\begin{aligned}
A_0 &= \{O_1^{-1} O_2^{-1} O_3^{-1}\} \\
A_1 &= \{O_1 O_2^{-1} O_3^{-1}, O_1^{-1} O_2 O_3^{-1}, O_1^{-1} O_2^{-1} O_3\} \\
A_2 &= \{O_1^{-1} O_2 O_3, O_1 O_2^{-1} O_3, O_1 O_2 O_3^{-1}\} \\
A_3 &= \{O_1 O_2 O_3\}.
\end{aligned}$$

Hence,

$$\begin{aligned}
\Pr\{J = 0\} &= \Pr\{O_1^{-1} O_2^{-1} O_3^{-1}\} \\
&= 1 - \psi_1(1) - \psi_1(2) - \psi_1(3) + \psi_2(1,2) + \psi_2(1,3) \\
&\quad + \psi_2(2,3) - \psi_3(1,2,3), \\
\Pr\{J = 1\} &= \Pr\{O_1 O_2^{-1} O_3^{-1}\} + \Pr\{O_1^{-1} O_2 O_3^{-1}\} + \Pr\{O_1^{-1} O_2^{-1} O_3\} \\
&= \psi_1(1) - \psi_2(1,2) - \psi_2(1,3) + \psi_3(1,2,3) \\
&\quad + \psi_1(2) - \psi_2(1,2) - \psi_2(2,3) + \psi_3(1,2,3) \\
&\quad + \psi_1(3) - \psi_2(1,3) - \psi_2(2,3) + \psi_3(1,2,3), \\
\Pr\{J = 2\} &= \Pr\{O_1^{-1} O_2 O_3\} + \Pr\{O_1 O_2^{-1} O_3\} + \Pr\{O_1 O_2 O_3^{-1}\} \\
&= \psi_2(2,3) - \psi_3(1,2,3) \\
&\quad + \psi_2(1,3) - \psi_3(1,2,3) \\
&\quad + \psi_2(1,2) - \psi_3(1,2,3),
\end{aligned}$$

and

$$\Pr\{J = 3\} = \psi_3(1,2,3).$$

Example 5.4. Consider a trapezoidal region, C, with dimensions $u = 20$[m] a[illegible] $w = 60$[m]. The disks are scattered according to a bivariate normal distribution, wi[illegible] location at $\xi = 0$[m], $\eta = 40$[m], $\sigma_x = \sigma_y = 30$[m], $\rho = .5$. The intensity over C [illegible] $\mu = 25$. Moreover, the radii of disks are uniformly distributed over $[0, 1]$. There a[illegible] $m = 5$ targets at orientations: 25°, 15°, 5°, −5° and −15°. Using program VPNRMA[illegible] we obtain the following visibility probabilities:

Table 5.1. Visibility Probabilities

i	j	k	l	m	ψ	i	j	k	l	m	ψ
1					.8506	1	2	3			.5899
2					.8355	1	2	4			.5915
3					.8302	1	2	5			.5980
4					.8325	1	3	4			.5878
5					.8414	1	3	5			.5941
1	2				.7106	1	4	5			.5959
1	3				.7061	2	3	4			.5773
1	4				.7081	2	3	5			.5835
1	5				.7156	2	4	5			.5853
2	3				.6935	3	4	5			.5816
2	4				.6954	1	2	3	4		.4911
2	5				.7030	1	2	3	5		.4963
3	4				.6911	1	2	4	5		.4978
3	5				.6984	1	3	4	5		.4947
4	5				.7006	2	3	4	5		.4859
1	2	3			.5899	1	2	3	4	5	.4133

From these visibility probabilities we obtain the following probabilities of elementary events

Table 5.2. Probabilities of Elementary Events

j_1	j_2	j_3	j_4	j_5	P	j_1	j_2	j_3	j_4	j_5	P
1	1	1	1	1	.4133	-1	-1	-1	1	1	.0029
-1	1	1	1	1	.0726	-1	-1	1	-1	1	.0029
1	-1	1	1	1	.0814	-1	-1	1	1	-1	.0028
1	1	-1	1	1	.0845	-1	1	-1	-1	1	.0029
1	1	1	-1	1	.0830	-1	1	-1	1	-1	.0028
1	1	1	1	-1	.0778	-1	1	1	-1	-1	.0028
-1	-1	1	1	1	.0143	1	-1	-1	-1	1	.0032
-1	1	-1	1	1	.0149	1	-1	-1	1	-1	.0032
-1	1	1	-1	1	.0164	1	-1	1	-1	-1	.0031
-1	1	1	1	-1	.0136	1	1	-1	-1	-1	.0031
1	-1	-1	1	1	.0149	-1	-1	-1	-1	1	.0006
1	-1	1	-1	1	.0164	-1	-1	-1	1	-1	.0005
1	-1	1	1	-1	.0153	-1	-1	1	-1	-1	.0005
1	1	-1	-1	1	.0172	-1	1	-1	-1	-1	.0007
1	1	-1	1	-1	.0159	1	-1	-1	-1	-1	.0007
1	1	1	-1	-1	.0158	-1	-1	-1	-1	-1	.0000

nally, the above probabilities of elementary events yield the following distribution of number of visible targets, J:

Table 5.3. Distribution of The Number of Visible Targets

j	$\Pr\{J = j\}$	$\Pr\{J \leq j\}$
0	0.0000	0.0000
1	0.0030	0.0030
2	0.0297	0.0327
3	0.1547	0.1874
4	0.3993	0.5867
5	0.4133	1.0000

Accordingly, the probability of observing at least three targets is $\Pr\{J \geq 3\} = .9673$, and the probability of observing at most three targets is $\Pr\{J \leq 3\} = .1874$. Moreover, the median of the distribution of J is $M_e = 4$, and its expected value and standard-deviation are: $E\{J\} = 4.1902$, s.d.$\{J\} = .8232$. ■

We present now a general theory and method for computing the probabilities of elementary events.

5.1.2. General Method For Computing Probabilities of Elementary Events

Consider N lines of sight, $\mathcal{L}_1, \cdots, \mathcal{L}_N$, from ν observation points ($\nu \geq 1$) to m target points ($m \geq 1$), $N = \nu \cdot m$. Let $\mathcal{B} = \{B_j : j = 0, \cdots, 2^N - 1\}$ be the collection of all possible subsets of indices from $1, 2, \cdots, N$, where $B_0 \equiv \emptyset$, and $j = \sum_{n=1}^{N} b_{jn} 2^{n-1}$, where

$$b_{jn} = \begin{cases} 1, & \text{if } n \in B_j \\ 0, & \text{otherwise.} \end{cases} \tag{5.1}$$

These subsets of indices correspond to subsets of lines of sight. Define the indicator variables,

$$I_i = \begin{cases} 1, & \text{if } I_i \text{ is not intersected} \\ 0, & \text{otherwise.} \end{cases} \tag{5.2}$$

The elementary event corresponding to the subset B_j is

$$E_j = \{(I_1, I_2, \cdots, I_N) : I_n = b_{jn}, n = 1, \cdots, N\}, \quad j = 0, \cdots, 2^N - 1. \tag{5.3}$$

Notice that B_j is the set of line indices for which $I_n = 1$ in E_j.

Let $\psi\{B_j\}$ be the probability that all the lines with indices in B_j are simultaneously not intersected by random disks. $\psi\{B_j\}$ are the simultaneous visibility probabilities computed in the previous sections. We define $\psi\{B_0\} \equiv 1$. Let $\pi(E_j)$, $j = 0, \cdots, 2^N -$ denote the probability of E_j. Obviously, $\sum_{j=0}^{2^N-1} \pi(E_j) = 1$. Let $\boldsymbol{\pi}^{(N)}$ be a 2^N dimension probability vector whose j-th element is $\pi_j = \pi(E_j)$, $j = 0, \cdots, 2^N - 1$. Similarly, $\boldsymbol{\psi}^{(N)}$ be a 2^N dimensional vector whose j-th component is $\psi_j = \psi\{B_j\}$. In the previo section we provided several examples in which the probabilities of elementary events, π_j, were computed as linear functions of ψ_j. In the present section we show that, generally,

$$\pi^{(N)} = H_N \psi^{(N)}. \tag{5.4}$$

H_N is a 2^N by 2^N non-singular matrix, determined recursively as

$$H_1 = \begin{bmatrix} 1 & -1 \\ 0 & 1 \end{bmatrix}, \tag{5.5}$$

and, for $n = 2, \cdots, N$,

$$\begin{aligned} H_n &= H_1 \otimes H_{n-1} \\ &= \begin{bmatrix} H_{n-1} & -H_{n-1} \\ 0 & H_{n-1} \end{bmatrix}, \end{aligned} \tag{5.6}$$

where $\otimes$ denotes the Kronecker direct multiplication.

Relationship (5.4) can be proven by induction on N. One can easily verify Eq. (5.4) or $N = 1$, $N = 2$. Assuming that Eq. (5.4) holds fro all $N = 1, 2, \cdots, N^*$ we can show hat it holds also for $N^* + 1$. Indeed, let $E_j^{(N^*)} = \{(I_1, \cdots, I_{N^*}) : \sum_{n=1}^{N^*} I_n 2^{n-1} = j\}$ and $E_l^{(N^*+1)} = \{(I_1, \cdots, I_{N^*+1}) : \sum_{n=1}^{N^*+1} I_n 2^{n-1} = l\}$. Notice that if $l = 0, \cdots, 2^{N^*} - 1$,

$$E_l^{(N^*+1)} = (E_l^{(N^*)}, 0), \tag{5.7}$$

nd

$$E_{l+2^{N^*}}^{(N^*+1)} = (E_l^{(N^*)}, 1). \tag{5.8}$$

urthermore, for all $l = 0, \cdots, 2^{N^*} - 1$,

$$\begin{aligned} \pi_l^{(N^*+1)} &= \Pr\{E_l^{(N^*+1)}\} \\ &= \pi_l^{(N^*)} - \Pr\{(E_l^{(N^*)}, 1)\}. \end{aligned} \tag{5.9}$$

the induction hypothesis, $\pi^{(N^*)} = H_{N^*} \psi^{(N^*)}$. Moreover,

$$\psi^{(N^*+1)\prime} = (\psi^{(N^*)\prime}, \hat{\psi}^{(N^*)\prime}) \tag{5.10}$$

ere the vector $\hat{\psi}^{(N^*)\prime} = (\psi_{2^{N^*}}^{(N^*+1)}, \cdots, \psi_{2^{N^*+1}-1}^{(N^*+1)})$ and where

$$\psi_{j+2^{N^*}}^{(N^*+1)} = \psi\{B_j \cup \{N^* + 1\}\}, \quad j = 0, \cdots, 2^{N^*} - 1.$$

cordingly, for each $l = 0, \cdots, 2^{N^*} - 1$,

$$\Pr\{(E_l^{N^*}, 1)\} = H_l^{(N^*)} \hat{\psi}^{(N^*)}, \tag{5.11}$$

where $H_l^{(N^*)}$ is the l-th row of N^*. Finally, from Eqs. (5.8), (5.9) and (5.11) we obtain

$$\pi^{(N^*+1)} = \begin{bmatrix} H_{N^*} & -H_{N^*} \\ 0 & H_{N^*} \end{bmatrix} \psi^{(N^*+1)}$$
$$= H_{N^*+1}\psi^{(N^*+1)}.$$

Example 5.5. Consider the case of $N = 3$. The simultaneous visibility probabilities are

$$\psi_0 = 1,\ \ \psi_1 = \psi\{1\},\ \ \psi_2 = \psi\{2\},\ \ \psi_3 = \psi\{1,2\}$$
$$\psi_4 = \psi\{3\},\ \ \psi_5 = \psi\{1,3\},\ \ \psi_6 = \psi\{2,3\},\ \ \psi_7 = \psi\{1,2,3\}.$$

The matrix H_3 is

$$H_3 = \begin{bmatrix} 1 & -1 & -1 & 1 & -1 & 1 & 1 & -1 \\ 0 & 1 & 0 & -1 & 0 & -1 & 0 & 1 \\ 0 & 0 & 1 & -1 & 0 & 0 & -1 & 1 \\ 0 & 0 & 0 & 1 & 0 & 0 & 0 & -1 \\ 0 & 0 & 0 & 0 & 1 & -1 & -1 & 1 \\ 0 & 0 & 0 & 0 & 0 & 1 & 0 & -1 \\ 0 & 0 & 0 & 0 & 0 & 0 & 1 & -1 \\ 0 & 0 & 0 & 0 & 0 & 0 & 0 & 1 \end{bmatrix}.$$

Accordingly,

$$\pi_0 = 1 - \psi\{1\} - \psi\{2\} + \psi\{1,2\} - \psi\{3\} + \psi\{1,3\}$$
$$+\ \psi\{2,3\} - \psi\{1,2,3\},$$
$$\pi_1 = \psi\{1\} - \psi\{1,2\} - \psi\{1,3\} + \psi\{1,2,3\},$$
$$\pi_2 = \psi\{2\} - \psi\{1,2\} - \psi\{2,3\} + \psi\{1,2,3\},$$
$$\pi_3 = \psi\{1,2\} - \psi\{1,2,3\},$$
$$\pi_4 = \psi\{3\} - \psi\{1,3\} - \psi\{2,3\} + \psi\{1,2,3\},$$
$$\pi_5 = \psi\{1,3\} - \psi\{1,2,3\},$$
$$\pi_6 = \psi\{2,3\} - \psi\{1,2,3\},$$
$$\pi_7 = \psi\{1,2,3\}.$$

Example 5.6. In the present example we illustrate the procedure developed in t present section. Suppose that there are $n = 3$ target points and $m = 2$ observati points. Thus, there are $N = n \cdot m = 6$ possible lines of sight, and $2^6 = 64$ possil visibility sets B_j, and 64 elementary events. Suppose that the target points are locat at $\mathbf{T}_1 = (-10, 100)$[m], $\mathbf{T}_2 = (0, 100)$[m], $\mathbf{T}_3 = (10, 100)$[m]. The observation poi are at $\mathbf{O}_1 = (-5, 0)$[m] and $\mathbf{O}_2 = (5, 0)$[m]. The Poisson field is standard, and center at a strip $\mathcal{S}$ located between $u = 40$[m] and $w = 60$[m]. The intensity of the fi is $\lambda = 0.005[1/\text{m}^2]$. The radii of disks are uniformly distributed on $(0, 1)$. We a require that a whole interval (window) of length $L = 2$[m] will be completely observa around each target. With the aid of program VPWALL one can compute the vec

ψ of all the visibility probabilities. The 64 visibility probabilities ψ_j for the present example are given in Table 5.4. The visibility probabilities are arranged in standard order. The first one in the list is $\psi_0 \equiv 1$. The index of the set B_j is one minus the order index in the table. Thus, for example, if we wish to find the simultaneous visibility probability of the windows around the lines $\mathcal{L}_1$, $\mathcal{L}_2$ and $\mathcal{L}_6$ we consider the binary number $(1,1,0,0,0,1) = 1+2+2^5 = 35$. From the table $\psi_{35} = 0.5486$. The probabilities π_j $(j = 0, \cdots, 2^N - 1)$ of elementary events are given in Table 5.5. The computation of these values was performed according to Eq. (5.4), using the GAUSS® software. Thus, from Table 5.5, the probability of the elementary event that the windows around $\mathbf{T}_1$, $\mathbf{T}_2$ are observable from $\mathbf{O}_1$ and around $\mathbf{T}_2$, $\mathbf{T}_3$ are observable from $\mathbf{O}_2$, that around $\mathbf{T}_3$ is unobservable from $\mathbf{O}_1$ and around $\mathbf{T}_1$ is unobservable from $\mathbf{O}_3$, $\pi_{51} = 0.00377$. ■

Table 5.4. Visibility Probabilities Arranged In Standard Order

1	1.00000	17	0.81862	33	0.81862	49	0.67015
2	0.81862	18	0.67015	34	0.67015	50	0.54860
3	0.81862	19	0.67015	35	0.67015	51	0.54860
4	0.67015	20	0.54860	36	0.54860	52	0.44910
5	0.81781	21	0.74676	37	0.66948	53	0.61131
6	0.66948	22	0.61131	38	0.54805	54	0.50044
7	0.66948	23	0.61131	39	0.54805	55	0.50044
8	0.54805	24	.050044	40	0.44865	56	0.40967
9	0.81781	25	0.66948	41	0.66948	57	0.54805
10	0.66948	26	0.54805	42	0.54805	58	0.44865
11	0.74676	27	0.61131	43	0.61131	59	0.50044
12	0.61131	28	0.50044	44	0.50044	60	0.40967
13	0.66936	29	.061120	45	0.54795	61	0.50034
14	0.54795	30	0.50034	46	0.44857	62	0.40959
15	0.61120	31	0.55810	47	0.50034	63	0.45687
16	0.50034	32	0.45687	48	0.40959	64	0.37401

Table 5.5. Probabilities of Elementary Visibility Events

1	0.00052	17	0.00020	33	0.00175	49	0.00121
2	0.00175	18	0.00121	34	0.00819	50	0.00533
3	0.00020	19	0.00022	35	0.00121	51	0.00082
4	0.00121	20	0.00082	36	0.00533	52	0.00377
5	0.00021	21	0.00274	37	0.00119	53	0.01221
6	0.00119	22	0.01221	38	0.00523	54	0.05519
7	0.00019	23	0.00173	39	0.00074	55	0.00791
8	0.00074	24	0.00791	40	0.00340	56	0.03566
9	0.00021	25	0.00019	41	0.00119	57	0.00074
10	0.00119	26	0.00074	42	0.00523	58	0.00340

Table 5.5. Probabilities of Elementary Visibility Events (Continued)

11	0.00274	27	0.00173	43	0.01221	59	0.00791
12	0.01221	28	0.00791	44	0.05519	60	0.03566
13	0.00018	29	0.00174	45	0.00074	61	0.00789
14	0.00074	30	0.00789	46	0.00340	62	0.03558
15	0.00174	31	0.01837	47	0.00789	63	0.08286
16	0.00789	32	0.08286	48	0.03558	64	0.37401

5.1.3. Joint Distributions of Counting Variables

Consider the case of ν observation points and m target points. Let J_n denote the number of target points which are simultaneously observed from the n-th observation point $\mathbf{O}_n$, $n = 1, \cdots, \nu$. We are interested in the joint distribution of $(J_1, \cdots, J_\nu)$. In the previous section we established a procedure for determining all the 2^N probabilities of the elementary events, where $N = \nu \cdot m$. As before, let $I_n = 1$ if the line $\mathcal{L}_n$ is not intersected and $I_n = 0$, otherwise. Partition the set of indices $\{1, 2, \cdots, N\}$ to ν subsets $C_i = \{(i-1)m + 1, \cdots, im\}$, $i = 1, \cdots, \nu$. Let $\chi\{A\}$ denote the indicator variable, which assumes the value 1 if A is true and the value 0 if A is false. Then, the marginal p.d.f. of J_i $(i = 1, \cdots, \nu)$ is given by

$$p_i^{(N)}(k) = \sum_{l=0}^{2^N-1} \chi\left\{\sum_{j=1}^{N} I_j 2^{j-1} = l, \sum_{j \in C_i} I_j = k\right\} \pi_l^{(N)}, \tag{5.12}$$

where $\pi_l^{(N)} = \Pr\{E_l^{(N)}\}$ is the probability of the l-th elementary event. The joint p.d.f of J_i and $J_{i'}$, $i \neq i'$, is given by

$$p_{ii'}^{(N)}(k, k') = \sum_{l=0}^{2^N-1} \chi\left\{\sum_{j=1}^{N} I_j 2^{j-1} = l, \sum_{j \in C_i} I_j = k, \sum_{j \in C_{i'}} I_j = k'\right\} \pi_l^{(N)}. \tag{5.13}$$

In a similar manner we can obtain joint distributions of any subset of $(J_1, \cdots, J_\nu)$. In the following example we illustrate these distributions.

Example 5.7. We continue with the same set-up of Example 5.6. Let N_1 be the number of observable target points (with the windows) from $\mathbf{O}_1$, and let N_2 be that from $\mathbf{O}_2$. From the probabilities of elementary events, given in Table 5.5, we compute the following joint probability distribution of (N_1, N_2).

Table 5.6. Joint Probability Distribution of (N_1, N_2).

	N_2				
N_1	0	1	2	3	sum
0	0.00052	0.00216	0.00259	0.00074	0.00601
1	0.00216	0.01887	0.04075	0.01920	0.08098
2	0.00259	0.04075	0.16752	0.15410	0.36496
3	0.00074	0.01920	0.15410	0.37401	0.54805
sum	0.00601	0.08098	0.36496	0.54805	1.00000

According to this joint distribution, the expected values of N_i are $E\{N_1\} = E\{N_2\} = 2.46$, the variances are $V\{N_1\} = V\{N_2\} = 0.446$. Moreover, $\text{cov}(N_1, N_2) = 0.155$ and the coefficient of correlation is $\rho = 0.348$. ■

5.2. An Integrated Measure of Visibility on a Star-Shaped Curve

A curve $\mathcal{C}$ in a plane is called star-shaped if each ray from the origin, O, intersects $\mathcal{C}$ at most once. We consider now a region of obscuring disks, $\mathcal{S}$, and a star-shaped curve $\mathcal{C}$, which is entirely outside $\mathcal{S}$ such that, each ray from $\mathbf{O}$ intersecting $\mathcal{C}$ has to pass through $\mathcal{S}$ (see Figure 5.1.).

We assume that random disks centered in $\mathcal{S}$ cannot intersect $\mathbf{O}$ or $\mathcal{C}$.

Random disks on $\mathcal{S}$ cast their shadows on $\mathcal{C}$. These shadows are segments of $\mathcal{C}$ which are invisible from $\mathbf{O}$. In the present section we study an integrated measure of visibility, which is the total length of visible segments of $\mathcal{C}$.

Points on $\mathcal{C}$ are represented by their polar coordinates $(\rho(s), s)$, where s is the orientation angle and $\rho(s)$ the distance of a point from $\mathbf{O}$. We assume that the curve is piece-wise smooth, i.e., continuous derivatives of $\rho'(s)$ exist at almost all points s in (θ_L, θ_U).

The length of $\mathcal{C}$ between the points $\mathbf{P}_L$ and $\mathbf{P}_U$ is

$$L = \int_{\theta_L}^{\theta_U} [\rho^2(s) + (\rho'(s))^2]^{1/2} ds. \tag{5.14}$$

For example, if $\mathcal{C}$ is a segment of a circle of radius r, centered at $\mathbf{O}$, between the orientation angles θ_L and θ_U, $-\frac{\pi}{2} \leq \theta_L < \theta_U \leq \frac{\pi}{2}$, then each point on $\mathcal{C}$ is given by (r, s), $\theta_L \leq s \leq \theta_U$. The length of $\mathcal{C}$ is obviously $L = r(\theta_U - \theta_L)$. If $\mathcal{C}$ is a segment on a straight line of distance r from $\mathbf{O}$, and suppose that $\rho(0) = r$, then $\rho(s) = r/\cos(s)$. Thus, in the present case, points of $\mathcal{C}$ are represented by $(\frac{r}{\cos(s)}, s)$, $\theta_L \leq s \leq \theta_U$. In this case $\rho'(s) = r\ \sin(s)/\cos^2(s)$, $(\rho^2(s) + (\rho'(s))^2)^{1/2} = r/\cos^2(s)$ and the length of $\mathcal{C}$ is $L = r\int_{\theta_L}^{\theta_U} ds/\cos^2(s) = r(\tan(\theta_U) - \tan(\theta_L))$.

The total length of the segments on $\mathcal{C}$ which are visible is given by

$$V = \int_{\theta_L}^{\theta_U} I(s)[\rho^2(s) + (\rho'(s))^2]^{1/2} ds, \tag{5.15}$$

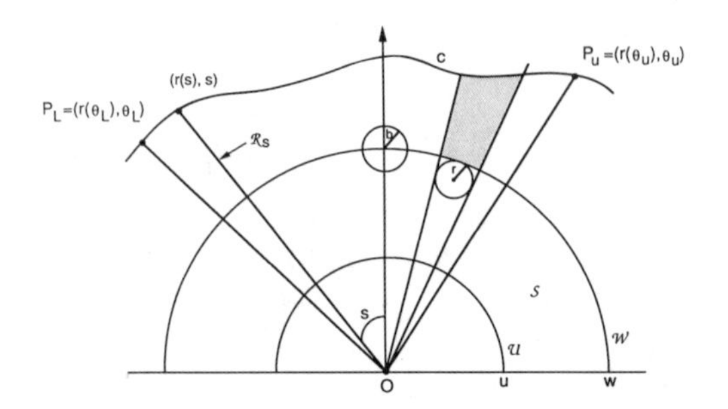

Figure 5.1. The Geometry of Disks Casting Shadows on $\mathcal{C}$, With Annular Region $\mathcal{S}$

where $I(s) = 1$ if the point $(\rho(s), s)$ on $\mathcal{C}$ is visible, and $I(s) = 0$ otherwise. V is a random variable, which depends on the random scattering of the obscuring disks and on their random radii. We consider the proportion of the visibility measure $W = V/L$. This proportional visibility measure can be written as

$$W = \int_{\theta_L}^{\theta_U} I(s)l(s)ds, \tag{5.16}$$

where $l(s)$ is a p.d.f. on $[\theta_L, \theta_U]$, given by

$$l(s) = [\rho^2(s) + (\rho'(s))^2]^{1/2}/L, \quad \theta_L \leq s \leq \theta_U. \tag{5.17}$$

One could generalize the definition of W by using any p.d.f. $l(s)$ concentrated on the interval (θ_L, θ_U), not necessarily $l(s)$ of Eq. (5.17). Such a function satisfies $l(s) \geq 0$ on (θ_L, θ_U), $l(s) = 0$ for $s \notin (\theta_L, \theta_U)$ and $\int_{\theta_L}^{\theta_U} l(s)ds = 1$. Notice that $W = 0$ if the whole segment $\mathcal{C}$ is invisible from $\mathbf{O}$ (in a shadow). Let $p_0 = \Pr\{W = 0\}$. $W = 1$ if the whole segment $\mathcal{C}$ is visible. Let $p_1 = \Pr\{W = 1\}$. The value of p_1 can be computed by the methods of the previous chapters. Specific formula for p_1 will be given later for a special case. The determination of p_0 is much more difficult. Approximations for p_0 will be discussed later in this chapter. In Chapter 6 we will study the problem of the distribution of shadows length on $\mathcal{C}$ and will discuss also the problem of determining the value of p_0. Let $H_W(w)$ denote the c.d.f. of W. Thus,

$$H_W(w) = \begin{cases} 0, & \text{if } w < 0 \\ p_0, & \text{if } w = 0 \\ H_W^*(w), & \text{if } 0 < w < 1 \\ 1, & \text{if } 1 \leq w, \end{cases} \tag{5.18}$$

where $p_1 = 1 - H_W^*(1)$. A beta approximation to $H_W^*(w)$ will be discussed later. From Eq. (5.18) we deduce that the n-th moment of W is

$$E\{W^n\} = p_1 + \int_0^1 w^n dH_W^*(w). \tag{5.19}$$

In the next section we develop an algorithm for the computation of the moments $E\{W^n\}$ as functionals of visibility probabilities. Notice that Eq. (5.19) implies that

$$\lim_{n\to\infty} E\{W^n\} = p_1. \tag{5.20}$$

5.3. The Moments of W

Let $\psi_1(s)$ designate the visibility probability of the point $(\rho(s), s)$ on $\mathcal{C}$. This visibility probability can be computed by the methods discussed in the previous chapters. Since $E\{I(s)\} = \psi_1(s)$, the expected value of the proportional visibility measure W is

$$\begin{aligned} E\{W\} &= \int_{\theta_L}^{\theta_U} E\{I(s)\}l(s)ds \\ &= \int_{\theta_L}^{\theta_U} \psi_1(s)l(s)ds. \end{aligned} \tag{5.21}$$

Generally, if $\psi_n(s_1, \cdots, s_n)$ designates the simultaneous visibility probability of n points on $\mathcal{C}$, where $\theta_L \le s_n < s_{n-1} < \cdots < s_1 \le \theta_U$, then the n-th moment of W is given by,

$$\begin{aligned} E\{W^n\} &= \int_{\theta_L}^{\theta_U} \cdots \int_{\theta_L}^{\theta_U} E\left\{\prod_{i=1}^n I(s_i)\right\} \prod_{i=1}^n l(s_i)ds_i \\ &= n! \int_{\theta_L}^{\theta_U} l(s_1)ds_1 \int_{\theta_L}^{s_1} l(s_2)ds_2 \cdots \int_{\theta_L}^{s_{n-1}} l(s_n)\psi_n(s_1, \cdots, s_n)ds_n. \end{aligned} \tag{5.22}$$

From Eq. (5.21) we conclude that $E\{W\}$ is the expected visibility probability $\psi_1(s)$, when the orientation coordinate s is chosen at random from the interval (θ_L, θ_U), according to the distribution with p.d.f. $l(s)$. Similarly, Eq. (5.22) implies that $E\{W^n\}$ the expected value of $\psi_n(s_1, \cdots, s_n)$, where $\theta_L \le s_n < \cdots < s_1 \le \theta_U$ is the order statistic of n independent random variables having an identical distribution with p.d.f. $l(s)$. From Eqs. (3.48)-(3.50) we obtain the formula

$$\begin{aligned} E\{W^n\} = n!e^{-\mu} &\int_{\theta_L}^{\theta_U} \exp\{\mu K_+(s_1, \theta_U - s_1)\}l(s_1) \cdot \\ &\int_{\theta_L}^{s_1} \exp\left\{\mu\left[K_-\left(s_1, \frac{s_1 - s_2}{2}\right) + K_+\left(s_2, \frac{s_1 - s_2}{2}\right)\right]\right\} l(s_2) \cdots \\ &\cdots \int_{\theta_L}^{s_{n-1}} \exp\left\{\mu\left[K_+\left(s_n, \frac{s_{n-1} - s_n}{2}\right) + K_-(s_n, s_n - \theta_L^*)\right]\right\} l(s_n) \\ &ds_n ds_{n-1} \cdots ds_1, \quad n \ge 2 \end{aligned} \tag{5.23}$$

where $\mu = \lambda A\{C\}$, and

$$E\{W\} = e^{-\mu} \int_{\theta_L}^{\theta_U} \exp\{\mu K_+(s, \theta_U^* - s)\} \exp\{\mu K_-(s, s - \theta_L^*)\} l(s) ds.$$

If the disks are centered in an annular region between $\mathcal{U}$ and $\mathcal{W}$, of distance u and w from $\mathbf{O}$ and their radii are in (a, b) then θ_L^* and θ_U^* are given by Eq. (5.31). Define

$$G_0(s) = \exp\{\mu K_-(s, s - \theta_L^*)\},$$

and (5.24)

$$G^*(s) = \exp\{\mu K_+(s, \theta_U^* - s)\}.$$

Then, we can write

$$E\{W\} = e^{-\mu} \int_{\theta_L}^{\theta_U} G_0(s) G^*(s) l(s) ds. \tag{5.25}$$

For $n = 2$ we obtain from (5.23)-(5.25),

$$E\{W^2\} = 2e^{-\mu} \int_{\theta_L}^{\theta_U} G^*(s_1) l(s_1) \int_{\theta_L}^{s_1} G_0(s_2) \exp\left\{\mu \left[K_-\left(s_1, \frac{s_1 - s_2}{2}\right) + K_+\left(s_2, \frac{s_1 - s_2}{2}\right)\right]\right\} l(s_2) ds_2 ds_1. \tag{5.26}$$

Let

$$G_1(s) = \int_{\theta_L}^{s} G_0(y) \exp\left\{\mu \left[K_-\left(s, \frac{s-y}{2}\right) + K_+\left(y, \frac{s-y}{2}\right)\right]\right\} l(y) dy \tag{5.27}$$

then

$$E\{W^2\} = 2e^{-\mu} \int_{\theta_L}^{\theta_U} G_1(s) G^*(s) l(s) ds. \tag{5.28}$$

Generally, define recursively, for every $j \geq 1$,

$$G_j(s) = \int_{\theta_L}^{s} G_{j-1}(y) \exp\left\{\mu \left[K_-\left(s, \frac{s-y}{2}\right) + K_+\left(y, \frac{s-y}{2}\right)\right]\right\} l(y) dy. \tag{5.29}$$

Then, the moments of W can be determined by the formula

$$E\{W^n\} = n! e^{-\mu} \int_{\theta_L}^{\theta_U} G_{n-1}(s) G^*(s) l(s) ds. \tag{5.30}$$

The computation of the moments according to these recursive equations can proceed by numerical integrations. Program MOTRAP performs these computations, for the case of a trapezoidal region. The output of the program are the required moments, as well as those of a mixed-beta approximation, which are discussed in the following section.

In the following table we present the first 5 moments of W, for a trapezoidal region with parameters $\theta_L = -\pi/18$, $\theta_U = \pi/18$, $r = 100$ [m], $u = 50$ [m], $w = 75$ [m]. The radii of the disks are uniformly distributed on $(0, b)$.

		Moment	of	Order		
b	λ	1	2	3	4	5
1.5	0.001	.9630	.9306	.9022	.8772	.8551
10	0.005	.2848	.1587	.1079	.0821	.0672

Another approach is to approximate the values of the moments by a Monte Carlo technique. n values of s, in the interval (θ_L, θ_U) are randomly chosen according to a distribution with p.d.f. $l(s)$. These randomly chosen s values are then ordered, and the visibility function $\psi_n(s_1, \cdots, s_n)$ computed. An average of many independent replicas of this yields an estimate of $E\{W^n\}$, which converges, with probability 1, to the true value. More formally, let $L(s)$ be the c.d.f. corresponding to $l(s)$, i.e.,

$$L(s) = \begin{cases} 0, & s < \theta_L \\ \int_{\theta_L}^{s} l(y)dy, & \theta_L \le s < \theta_U \\ 1, & \theta_U \le s. \end{cases}$$

Let $L^{-1}(p)$, for $0 < p < 1$, denote the p-th fractile of $L(s)$. Determine the number of replicas, Nr, to run. Let $U_1, U_2, \cdots$ be random numbers following a uniform distribution on $(0, 1)$. These numbers are generated by calling available computer routines. The following algorithm computes a Monte Carlo approximation to $E\{W^n\}$, $n = 1, 2, \cdots$

Step 0. Set n, Nr, $j = 1$, SUM $= 0$
Step 1. For $i = 1, \cdots, n$ compute $s_i = L^{-1}(U_i)$.
Step 2. Order the n s-values decreasingly $s_n < s_{n-1} < \cdots < s_1$
Step 3. Compute $\hat{\psi}_{n,j} = \psi_n(s_1, \cdots, s_n)$, SUM = SUM + $\hat{\psi}_{n,j}$.
Step 4. If $j = Nr$ then GOTO Step 5; Else, set $j \leftarrow j + 1$ GOTO Step 1.
Step 5. Compute $\hat{W}_{n,Nr} = \frac{1}{Nr}$SUM.

$\hat{W}_{n,Nr}$ is an estimate of the n-th moment $E\{W^n\}$. Program MOANNTSM computes the moments of W, in the case of annular region $\mathcal{S}$ and $\mathcal{C}$ on a concentric circle, by this Monte Carlo Method.

Example 5.7. We consider here an annular strip $\mathcal{S}$ and a concentric circle for $\mathcal{C}$. More specifically, let the boundaries of $\mathcal{S}$ be

$$\mathcal{U} = \left\{(u, s),\ -\frac{\pi}{2} \le s \le \frac{\pi}{2}\right\},$$
$$\mathcal{W} = \left\{(w, s),\ -\frac{\pi}{2} \le s \le \frac{\pi}{2}\right\}$$

and let

$$\mathcal{C} = \{(r, s),\ \theta_L \le s \le \theta_U\},$$

where $0 < u < w < r$.

Disk are scattered in $\mathcal{S}$ according to a standard Poisson random field, with radii having uniform distributions on $[a, b]$. We assume that $b < \min(u, r - w)$. In Section 3.2.1 we showed that in the present case $K_+(s,t) = K_-(s,t) = K^*(t,w) - K^*(t,u)$, where $K^*(t,v)$ is given by (3.56)-(3.58) and $A\{C\}$ is given by (3.59). Since $\mathcal{C}$ is located on a circle centered at $\mathbf{O}$ with radius r, $L = r(\theta_U - \theta_L)$ and $l(s) = 1/(\theta_U - \theta_L)$, for $\theta_L \le s \le \theta_U$. That is, $l(s)$ is a density of a uniform distribution on (θ_L, θ_U). In the following table we give the first six moments of W, computed according to the Monte Carlo approximation (Program MOANNTSM), with $Nr = 100$ replicas, for several values of λ [1/m^2].

The parameters are: $r = 100$[m], $u = 40$[m], $w = 60$[m], $a = 1$[m], $b = 2$[m], $\lambda = 0.001$[1/m^2], $\theta_L = -\pi/6$, $\theta_U = \pi/6$.

Table 5.6. Moments of W

λ	1	2	3	4	5	6
0.001	0.9418	0.8919	0.8545	0.8296	0.8119	0.8017
0.002	0.8869	0.8028	0.7363	0.6963	0.6717	0.6572
0.003	0.8352	0.7101	0.6279	0.5697	0.5353	0.5176
0.004	0.7866	0.6346	0.5342	0.4657	0.4275	0.4112
0.005	0.7408	0.5655	0.4596	0.4006	0.3641	0.3454

5.4. Approximations to the Distribution of W

5.4.1. A Beta Approximation.

Let p_1 denote the probability that the whole portion of $\mathcal{C}$, from $\mathbf{P}_L$ to $\mathbf{P}_U$, is visible Notice that, in order to assure that the points close to θ_L or θ_U be visible, we have t guarantee that obscuring disks are not centered to the left of θ_L, or to the right of θ_U and cast shadows on the portion of $\mathcal{C}$ under consideration. For this purpose we conside a point $\mathbf{P}'_L$ to the left of $\mathbf{P}_L$ and a point $\mathbf{P}'_U$ to the right of $\mathbf{P}_U$ having orientatio coordinates

$$\theta_L^* = \theta_L - \sin^{-1}\left(\frac{b}{u}\right), \quad \text{and} \quad \theta_U^* = \theta_U + \sin^{-1}\left(\frac{b}{u}\right). \qquad (5.31$$

If there are no disks centered between θ_L and θ_U and if those centered between θ and θ_L, or θ_U and θ_U^*, do not intersect the rays R_{θ_L} or R_{θ_U}, then the whole segment $\mathcal{C}$, between $\mathbf{P}_L$ and $\mathbf{P}_U$, is visible. Hence,

$$\begin{aligned} p_1 = \exp\{-\lambda A\{C\}\} \exp\{\mu[&K_-(\theta_L, \theta_L - \theta_L^*) \\ &+ K_+(\theta_U, \theta_U^* - \theta_U)]\}. \end{aligned} \qquad (5.3$$

Thus, the probability of complete visibility can be readily determined. Let p_0 be t probability that the segment of $\mathcal{C}$ of interest is completely invisible. One can appro imate the value of p_0 in several ways. One method of approximation, is to partiti the interval (θ_L, θ_U) to m equal parts. We consider then the set of $(m+1)$ points $\mathbf{P}_j$ $(j = 0, 1, \cdots, m)$ on $\mathcal{C}$, having orientation angles $s_j = \theta_L + j(\theta_U - \theta_L)/m$. We can then compute, by the method of Section 5.1, the elementary probability $\Pr\{O_0^{-1} O_1^{-1} \cdots O_m^{-1}\}$ that all the $(m+1)$ points are invisible. Obviously, $p_0 \leq \Pr\{O_0^{-1} O_1^{-1} \cdots O_m^{-1}\}$ for every $m \geq 1$. Moreover, $p_0 = \lim_{m \to \infty} \Pr\{O_0^{-1} \cdots O_m^{-1}\}$.

In Example 5.4 we computed all the 32 probabilities of the elementary events associated with 5 points. If those points are all within the segment of interest, then $p_0 \leq \Pr\{O_1^{-1} \cdots O_5^{-1}\}$, which was found to be smaller than 10^{-4}. Another possible approximation to p_0 is given by Eq. (5.39). As mentioned before, the c.d.f. of W has jump points p_0 and p_1 at $w = 0$ and $w = 1$, respectively. The distribution of W is generally absolutely continuous for values of W within $(0,1)$. Thus, the c.d.f. of W, $H_W(w)$, can be written as in Eq. (5.18) with

$$H_W(w) = p_0 + (1 - p_0 - p_1) \int_0^w h_W^*(x)dx, \tag{5.33}$$

where $h_W^*(w)$ is a p.d.f. over $(0,1)$.

A density function $f(x)$ on $(0,1)$ is a p.d.f. of a beta distribution, with parameters α, β, where $0 < \alpha, \beta < \infty$, if

$$f(x) = \frac{1}{B(\alpha, \beta)} x^{\alpha-1}(1-x)^{\beta-1}, \quad 0 \leq x \leq 1, \tag{5.34}$$

where

$$B(\alpha, \beta) = \int_0^1 x^{\alpha-1}(1-x)^{\beta-1} dx. \tag{5.35}$$

We approximate the c.d.f. of W, $H_W(w)$, by the beta-mixture

$$\tilde{H}_W(w) = p_0 + (1 - p_0 - p_1) \frac{1}{B(\alpha, \beta)} \int_0^w x^{\alpha-1}(1-x)^{\beta-1} dx, \quad 0 < w < 1. \tag{5.36}$$

In order to obtain values of α and β, which will yield an approximation to the true distribution of W, we apply the method of moments, and determine the values of p_0, α and β which will yield the same values to the first three moments of $H_W(w)$ and of $\tilde{H}(w)$. From the formula of the moments of a beta distribution we obtain that the n-th moment of $\tilde{H}(w)$ is

$$\tilde{\mu}_n = p_1 + (1 - p_0 - p_1) \frac{\alpha(\alpha+1) \cdots (\alpha+n-1)}{(\alpha+\beta)(\alpha+\beta+1) \cdots (\alpha+\beta+n-1)}. \tag{5.37}$$

Thus, if μ_n is the n-th moment of W, and $C_n = \mu_n - p_1$ then,

$$\begin{aligned} \alpha &= \frac{2C_2^2 - C_3(C_1 + C_2)}{C_1 C_3 - C_2^2} \\ \beta &= \frac{(C_1 - C_2)(C_2 - C_3)}{C_1 C_3 - C_2^2} \end{aligned} \tag{5.38}$$

and

$$p_0 = \left(1 - p_1 - \frac{C_1(\alpha+\beta)}{\alpha}\right)^+ . \tag{5.39}$$

Eq. (5.39) provides another approximation to the value of p_0. As long as the value of p_0 given by (5.39) is smaller than the upper bound discussed earlier then the beta-mixture approximation can be considered as a good one. The distribution of the right-hand limit of a shadow, given in Section 6.4, can be used to determine p_0 (see Problem [6.5.1]).

Example 5.8. In the present example we provide a numerical comparison of the moments of W to those of the beta-mixture distribution. We consider an annular region with $u = 50$[m], $w = 75$[m] and the curve $\mathcal{C}$ is a half circle centered at the origin, with radius $r = 100$[m]. The segment of $\mathcal{C}$ under consideration is between $\theta_L = -\pi/18$ and $\theta_U = \pi/18$ the radii of disks are uniformly distributed between $a = 5$[m] to $b = 15$[m]. The centers of disks are scattered according to a homogeneous Poisson field.

In the following table we present the first $n = 6$ moments of W, and the value of $p_1 = \mu_\infty$, together with their beta-mixture approximations, for various values of λ. The results in Tables 5.7 and 5.8 were computed with program MOANNTSM.

Table 5.7. Moments of W and Beta-Mixture Approximations$^{+)}$

Intensity	Moment			Order			
$\lambda/[10^4]$	1	2	3	4	5	6	∞
1	.951	.938	.931	.928	.927	.926	.901
	.951	.938	.931	.927	.924	.922	
3	.860	.833	.816	.808	.804	.802	.730
	.860	.833	.816	.803	.794	.787	
5	.778	.731	.706	.694	.687	.684	.592
	.778	.731	.706	.691	.680	.672	

$^{+)}$ The moments of W are at the upper line of each cell, and those of the beta-mixture approximation are at the lower line.

We see that the fourth, fifth and six moments of $H_W(w)$ and of $\tilde{H}(w)$ are very close. The parameters of the mixed-beta distributions $\tilde{H}(w)$ are

Table 5.8. Parameters of Mixed-Beta Distributions

λ	σ	p_0	p_1	α	β
1	.1821	.0000	.9005	.4352	.5230
3	.3058	.0934	.7302	2.7787	.9873
5	.3551	.0739	.5921	.7285	.5824

The value of σ is that of the standard deviation of W.

5.4.2. Discrete Approximation.

Let $L(s)$ denote the c.d.f. corresponding to the p.d.f. $l(s)$ on (θ_L, θ_U). Let $\bar{\mathcal{C}}$ be the line segment on $\mathcal{C}$, from $\mathbf{P}_{\theta_L}$ to $\mathbf{P}_{\theta_U}$. We partition $\bar{\mathcal{C}}$ to N subsegments of equal weight. The boundaries of these subsegments are

$$\xi_i = L^{-1}\left(\frac{i}{N}\right) = \inf\left\{s : s \geq \theta_L,\ L(s) \geq \frac{i}{N}\right\}, \tag{5.40}$$

for $i = 0, 1, \cdots, N$. Notice that $\xi_0 = \theta_L$ and $\xi_N = \theta_U$. In terms of this partition

$$W = \sum_{i=1}^{N} \int_{i=1}^{\xi_i} I(s) l(s) ds.$$

Let $\mathbf{P}_{s_i}$ be the midpoint of the i-th subsegment, i.e.,

$$s_i = L^{-1}\left(\frac{i - 1/2}{N}\right), \qquad i = 1, \cdots, N. \tag{5.41}$$

Around each mid-point $\mathbf{P}_{s_i}$ we define a segment $J_{N,\delta}(s_i)$, $i = 1, \cdots, N$, of weight $\frac{\delta}{N}$ $(0 \leq \delta \leq 1)$. That is,

$$J_{N,\delta}(s_i) = \{\mathbf{P}_s : s_i^-(\delta) \leq s \leq s_i^+(\delta)\}, \tag{5.42}$$

where

$$s_i^-(\delta) = L^{-1}\left(\frac{i}{N} - \frac{1+\delta}{2N}\right)$$

and (5.43)

$$s_i^+(\delta) = L^{-1}\left(\frac{i}{N} - \frac{1-\delta}{2N}\right).$$

Notice that when $\delta = 0$ then $s_i^-(0) = s_i = s_i^+(0)$, and when $\delta = 1$ then $s_i^-(1) = \xi_{i-1}$ and $s_i^+(1) = \xi_i$. Finally, let $I_{N,\delta}(s_i) = 1$ if $J_{N,\delta}(s_i)$ is *completely visible*, and $I_{N,\delta}(s_i) = 0$, otherwise. Define the discrete random variable

$$\begin{aligned} W_N(\delta) &= \sum_{i=1}^{N} \int_{\xi_{i-1}}^{\xi_i} I_{N,\delta}(s_i) l(s) ds \\ &= \frac{1}{N} \sum_{i=1}^{N} I_{N,\delta}(s_i). \end{aligned} \tag{5.44}$$

$W_N(\delta)$ is the relative frequency of $J_{N,\delta}$ segments on $\bar{\mathcal{C}}$, which are completely visible. Notice that if $\delta_1 < \delta_2$ then $W_N(\delta_1) \geq W_N(\delta_2)$ with probability one.

$W_N(\delta)$ converges uniformly to W, as $N \to \infty$, with probability one. Indeed, let $K(\bar{\mathcal{C}})$ be the number of shadows on $\bar{\mathcal{C}}$. $K(\bar{\mathcal{C}})$ is a random variable. Since the field of obscuring elements (disks) is Poisson, $K(\bar{C})$ is finite with probability one, and its distribution is independent of N and δ. Furthermore,

$$\begin{aligned}|W_N(\delta) - W| \leq \frac{1}{N}\sum_{i=1}^{N} \sup\{|I_{N,\delta}(s_i) \\ - I(s)| : \xi_{i-1} \leq s \leq \xi\} \leq \frac{2}{N}K(\bar{C}).\end{aligned} \tag{5.45}$$

Let $K_{1-\eta}$ be the $(1-\eta)$ fractile of $K(\overline{C})$, and let $N(\epsilon,\eta) = 2K_{1-\eta}/\epsilon$. Then

$$P\{|W_N(\delta) - W| < \epsilon, \quad \text{for all} \quad N \geq N(\epsilon,\eta)\} \geq 1-\eta.$$

This proves the uniform convergence of $W_N(\delta)$ to W with probability 1. Thus, we approximate the distribution of W, $H_W(w)$, by the c.d.f. of $W_N(\delta)$, which we denote by $H_W^{(N)}(w;\delta)$. Let $h^{(N)}(k;\delta) = \Pr\{W_N(\delta) = \frac{k}{n}\}$. Then,

$$H_W^{(N)}(w;\delta) = \begin{cases} 0, & \text{if } w < 0 \\ \sum_{j=0}^{k} h^{(N)}(j;\delta), & \text{if } \frac{k}{N} \leq w < \frac{k+1}{N},\ k = 0,1,\cdots,N-1 \\ 1, & \text{if } 1 \leq w. \end{cases}$$

Notice that $p_0 \leq h^{(N)}(0;\delta)$ and $h^{(N)}(1;1) = p_1$. Moreover, since $W \geq W_N(1)$ for all N, with probability 1, $H_W(w) \leq H^{(N)}(w;1)$ for all $N \geq 1$.

The computation of $h^{(N)}(k;\delta)$ can be done according to the methods discussed ir Sections 5.1.1-5.1.3. However, in order to obtain good approximation, the numbe of subsegments, N, should be large. In such a case it is impractical to compute al the 2^N elementary probabilities of the possible realizations of the values of $I_{N,\delta}(s_i)$ $i = 1,\cdots,N$. We provide below a dynamic algorithm developed by Yadin and Zack (1986) for the determination of the probability distribution $h^{(N)}(k;\delta)$. In the numerica example, which we later present, the values of N are 40 and 50.

5.4.3. Recursive Determination of $h^{(N)}(k;\delta)$.

We start by defining the following quantities. For $i = 1,\cdots,N$, let $\xi^-_{N,\delta}(i)$ [respectivel $\xi^+_{N,\delta}(i)$] be the probability that $J_{N,\delta}(s_i)$ is unshadowed by disks centered to the le [respectively, to the right] of the ray $\mathbb{R}_{s_i}$ (from $\mathbf{O}$ to $\mathbf{P}_{s_i}$).

If $\mathcal{S}$ is, for example a strip between two parallel lines $\mathcal{U}$, $\mathcal{W}$, and C is a line parallel t $\mathcal{S}$, and the distances of $\mathcal{U}$, $\mathcal{W}$ and C from $\mathbf{O}$ are $u < w < r$, respectively (the trapezo case) then, for a standard Poisson field,

$$\xi^-_{N,\delta}(i) = \exp\{-\lambda[T(\theta^*_L, s_i) - T(\theta^*_L,\theta^*_U)\cdot K_-(s_i^-(\delta), s_i^-(\delta) - \theta^*_L)]\} \tag{5.4}$$

where $\mathbf{P}_{\theta'}$ is a point left of $\mathbf{P}_{\theta_L}$ such that a disk centered left to the ray $\mathbb{R}_{\theta_L^*}$ cannot cast a shadow on $\bar{C}$. Furthermore,

$$T(\theta_1, \theta_2) = r(w^2 - u^2)(\tan(\theta_2) - \tan(\theta_1))/2. \tag{5.47}$$

Similarly,

$$\xi^+_{N,\delta}(i) = \exp\{-\lambda[T(s_i, \theta^*_U) - T(\theta^*_L, \theta^*_U) \cdot K_+(s^+_i(\delta), \theta^*_U - s^+_i(\delta))]\}. \tag{5.48}$$

The function $K_+(s,t)$ for the trapezoidal case is given in Eq. (3.64) - (3.66), and $K_-(s,t) = K_+(-s,t)$. For $1 \leq i < j \leq N$, let $\xi_{N,\delta}(i,j)$ designate the probability that both $J_{N,\delta}(s_i)$ and $J_{N,\delta}(s_j)$ are unshadowed by disks centered between the rays $\mathbb{R}_{s_i}$ and $\mathbb{R}_{s_j}$. This probability is given by the formula

$$\begin{aligned}\xi_{N,\delta}(i,j) = \exp\Bigg\{ &-\lambda\Bigg[T(s_i,s_j) - T(\theta^*_L,\theta^*_U)K_+\left(s^+_i(\delta), \frac{s^-_j(\delta) - s^+_i(\delta)}{2}\right) \\ &- T(\theta^*_L,\theta^*_U)K_-\left(s^-_j(\delta), \frac{s^-_j(\delta) - s^+_i(\delta)}{2}\right)\Bigg]\Bigg\}.\end{aligned} \tag{5.49}$$

Finally, for $n = 0, \cdots, N-1$; $m = 1, \cdots, N-n$ and $k = 0, \cdots, m-1$, let $Q_{N,\delta}(n,m,k)$ denote the probability that $J_{N,\delta}(s_n)$ *and* $J_{N,\delta}(s_{n+m})$ *and* exactly k out of the $m-1$ subsegments in between are unshadowed by disks centered between the rays $\mathbb{R}_{s_n}$ and $\mathbb{R}_{s_{n+m}}$.

From the above definitions we obtain,

$$Q_{N,\delta}(0,1,0) = \xi^-_{N,\delta}(1),$$

and (5.50)

$$Q_{N,\delta}(n,1,0) = \xi_{N,\delta}(n,n+1),$$

for $n = 1, \cdots, N-1$ and $Q_{N,\delta}(N,1,0) = \xi^+_{N,\delta}(N)$. Furthermore, from the renewal property of the semi-Markov process $\{I(s), \theta_L \leq s \leq \theta_u\}$,

$$Q_{N,\delta}(n,m,k) = \sum_{j=k}^{m-1} Q_{N,\delta}(n,j,k-1)Q_{N,\delta}(n+j,m-j,0), \tag{5.51}$$

for all $m = 2, \cdots, N-1$; $n = 0, \cdots, N-m$ and $k = 1, \cdots, m-1$. In addition

$$\sum_{k=0}^{m-1} Q_{N,\delta}(n,m,k) = \xi_{N,\delta}(n,n+m). \tag{5.52}$$

Hence,

$$Q_{N,\delta}(n,m,0) = \xi_{N,\delta}(n,n+m) - \sum_{k=1}^{m-1} Q_{N,\delta}(n,m,k). \tag{5.53}$$

Let $P_n(k)$, $n = 1, \cdots, N$ and $k = 0, \cdots, n$, denote the probability that k, out of the first n subsegments $J_{N,\delta}(s_1), \cdots, J_{N,\delta}(s_n)$, are completely visible. We compute these functions recursively according to the formulae

$$P_1(k) = \begin{cases} Q_{N,\delta}(0,1,0)\xi^+_{N,\delta}(1), & \text{if } k = 1 \\ \\ 1 - P_1(k), & \text{if } k = 0 \end{cases} \tag{5.54}$$

and

$$P_n(k) = \begin{cases} Q_{N,\delta}(0,n,n-1)\xi^+_{N,\delta}(n), & \text{if } k = n \\ \\ P_{n-1}(k) + [Q_{N,\delta}(0,n,k-1) - Q_{N,\delta}(0,n,k)]\xi^+_{N,\delta}(n), & \text{if } k = 1, \cdots, n-1 \\ \\ 1 - \sum_{j=1}^{n} P_j(j), & \text{if } k = 0. \end{cases} \tag{5.55}$$

Finally, the probability distribution function of $W_{N,\delta}$ is

$$h^{(N)}(k;\delta) = P_N(k), \qquad k = 0, 1, \cdots, N. \tag{5.56}$$

Example 5.9. Consider a linear curve with a trapezoidal region for disk centers. The radii disks are independent random variables having a uniform distribution on the interval $(0, 0.4)$. The path $\mathcal{C}$ is a horizontal straight line of distance $r = 100$ [m] from the origin. The horizontal boundaries of the strip $\mathcal{S}$ are at distances $u = 40$ [m] and $w = 60$ [m] from $\mathbf{O}$.

The segment $\bar{\mathcal{C}}$ is the interval between -100 [m] and 100 [m] on $\mathcal{C}$. The intensity of the Poisson field is λ [1/m^2]. In Table 5.9 we present the c.d.f. of $V_{N,\delta}$, for $N = 40, 50$, $\delta = 0, 1.0$ and $\mu = 10, 20$, where $\mu = \lambda T(-\pi/4, \pi/4)$. The c.d.f. $H_{N,\delta}(w)$ is tabulated for $w = 0(0.1)1^-$. $H_{N,\delta}(0)$ is the discrete approximation to the value of p_0. $1 - H_{N,1}(1-)$ is the discrete approximation to the value of p_1. In addition, we provide in Table 6.1 the mixed-beta approximation to the c.d.f., $\tilde{H}_W(x)$, given by Eq. (5.36). As seen in Table 5.9, $H_{50,1}(w) < H_{40,1}(w)$, for all $0 \le w < 1$. Indeed, $H_{N,1}(w)$ converges monotonically to $H(w)$ from above. Similarly, $H_{50,0}(0) < H_{40,0}(0)$ in all the two cases of $\mu = 10$ and 20. Indeed, $H_{N,0}(0) \ge p_0$ for all N, and $H_{N,0}(0) \to p_0$ as $N \to \infty$. We see there also that, for each λ, the values of the mixed-beta c.d.f. $\tilde{H}_W(w)$ satisfy

$$H_{50,0}(w) < \tilde{H}_W(w) < H_{50,1}(w), \qquad w = .2(.1).8.$$

This indicates that the mixed-beta approximation $\tilde{H}_W(w)$ is apparently close to the true c.d.f. $H_W(w)$, over a wide range of w. In Table 5.10 we compare the moments of $H_{N,0}(w)$, $H_{N,1}(w)$ and $\tilde{H}_W(w)$. We see that the moments of $H_{N,0}(w)$ are very close to those of $\tilde{H}_W(w)$. Recalling that the first three moments of $\tilde{H}_W(w)$ are, by definition the corresponding moments of W, we conclude that the moments of $H_{N,0}(w)$ are close to the true ones. The computations for Tables 5.9 and 5.10 were performed on a mainframe computer, due to the required memory size and amount of computations. A FORTRAN program can be furnished upon request.

Table 5.9. Values of $H_{N,\delta}(w)$ and of $\tilde{H}_W(w)$ for
$r = 100$ [m], $u = 40$ [m], $w = 60$ [m], $\theta' = \pi/4$, $\theta'' = \pi/4$, $\mu = 10, 20$, $b = .4$

		$\delta = 0$		$\delta = 1$		
μ	x	$N = 40$	$N = 50$	$N = 40$	$N = 50$	$H^*(x)$
	0.0	0.00081	0.00072	0.00225	0.00167	0.0000
10	0.1	0.00461	0.00501	0.01032	0.00962	0.0021
	0.2	0.02325	0.02461	0.04286	0.04036	0.0305
	0.3	0.07009	0.07315	0.11159	0.10657	0.0942
	0.4	0.15805	0.16336	0.22353	0.21624	0.1967
	0.5	0.29110	0.29864	0.37336	0.36519	0.3339
	0.6	0.45836	0.46724	.054266	0.53547	0.4954
	0.7	0.63657	0.64548	0.70731	0.70265	0.6645
	0.8	0.79584	0.80301	0.84177	0.83984	0.8190
	0.9	0.90577	0.91018	0.92824	0.92824	0.9325
	p_1	0.02620	0.02520	0.02140	0.02140	0.0214
	0.0	0.01588	0.01488	0.03489	0.02796	0.0084
20	0.1	0.06019	0.06440	0.10746	0.10287	0.0787
	0.2	0.18754	0.19516	0.27790	0.26839	0.2372
	0.3	0.36854	0.37822	0.47959	0.46829	0.4278
	0.4	0.56755	0.57417	0.66784	0.65805	0.6121
	0.5	0.73678	0.74463	0.81370	0.80708	0.7668
	0.6	0.86270	0.86802	0.90931	0.90579	0.8803
	0.7	0.93964	0.94263	0.96245	0.96102	0.9515
	0.8	0.97839	0.97974	0.98712	0.98672	0.9869
	0.9	0.99400	0.99449	0.99651	0.99646	0.9983
	p_1	0.00069	0.00054	0.00046	0.00046	0.0005

Table 5.10. Moments of $H_{N,\delta}(w)$ and of $\tilde{H}_W(w)$ for
$r = 100$ [m], $u = 40$ [m], $w = 60$ [m], $\theta' = \pi/4$, $\theta'' = \pi/4$, $\mu = 10, 20$, $b = .4$

		$\delta = 0$		$\delta = 1$		
	n	$N = 40$	$N = 50$	$N = 40$	$N = 50$	$H^*(x)$
$\mu = 10$	1	0.5933	0.5933	0.5488	0.5574	0.5933
	2	0.3951	0.3950	0.3477	0.3565	0.3949
	3	0.2843	0.2842	0.2416	0.2493	0.2840
	4	0.2167	0.2165	0.1794	0.1860	0.2158
	5	0.1726	0.1725	0.1402	0.1458	0.1710
	6	0.1425	0.1423	0.1140	0.1188	0.1400
	7	0.1209	0.1207	0.0957	0.0999	0.1177
	8	0.1051	0.1048	0.0825	0.0861	0.1011

Table 5.10. Moments of $H_{N,\delta}(w)$ and of $\tilde{H}_W(w)$ for
$r = 100$ [m], $u = 40$ [m], $w = 60$ [m], $\theta' = \pi/4$, $\theta'' = \pi/4$, $\mu = 10, 20$, $b = .4$
(Continued)

		δ	$= 0$	δ	$= 1$	
	n	$N = 40$	$N = 50$	$N = 40$	$N = 50$	$H^*(x)$
	9	0.0930	0.0928	0.0726	0.0758	0.0884
	10	0.0837	0.0834	0.0650	0.0679	0.0785
	1	0.3556	0.3556	0.3043	0.3139	0.3556
$\mu = 20$	2	0.1031	0.1630	0.1278	0.1340	0.1629
	3	0.0868	0.0867	0.0638	0.0677	0.0865
	4	0.0513	0.0512	0.0360	0.0385	0.0508
	5	0.0329	0.0329	0.0222	0.0239	0.0321
	6	0.0225	0.0225	0.0148	0.0160	0.0215
	7	0.0162	0.0162	0.0104	0.0113	0.0151
	8	0.0122	0.0122	0.0077	0.0084	0.0109
	9	0.0095	0.0095	0.0059	0.0064	0.0082
	10	0.0077	0.0076	0.0047	0.0051	0.0063

■

6

Distributions of Visible and Invisible Segments

In the present chapter we develop formulae for the distributions of the length of visible and invisible (shadowed) segments of the target curve $\mathcal{C}$ in the plane. We will focus attention on trapezoidal fields, in which the strip $\mathcal{S}$ is bounded by two parallel lines $\mathcal{U}$ and $\mathcal{W}$, and the target curve $\mathcal{C}$ is a straight line parallel to $\mathcal{S}$. As before, the distances of $\mathcal{U}$, $\mathcal{W}$, and $\mathcal{C}$ from $\mathbf{O}$ are u, w and r, respectively, where $0 < u < w < r$. We will assume that the Poisson field is standard and the distribution of the radius of a random disk centered in $\mathcal{S}$ is uniform on (a, b), and $2b < u$. The trapezoidal region is a subset of $\mathcal{S}$, C^*, bounded by $\mathcal{U}$, $\mathcal{W}$ and the rays $\mathbb{R}_{x_L^*}$ and $\mathbb{R}_{x_U^*}$, where x_L^* and x_U^* are the rectangular coordinates of points on $\mathcal{C}$ specified below. Let $\bar{\mathcal{C}}$ be an interval on $\mathcal{C}$ of interest. The rectangular x-coordinates of the points on $\bar{\mathcal{C}}$ are bounded by x_L and x_U, $x_L < x_U$, and, as before,

$$x_L^* = x_L - b\frac{r}{u}\left(1 + \left(\frac{x_L}{r}\right)^2\right)^{1/2}$$

$$x_U^* = x_U + b\frac{r}{u}\left(1 + \left(\frac{x_U}{r}\right)^2\right)^{1/2}.$$

We are interested in the distributions of the length of visible (in the light) segments of $\bar{\mathcal{C}}$ and that of the invisible (in shadow) segments. It is more convenient for the purposes of the present chapter to present points by their rectangular coordinates. The K-functions which we have used in the previous chapters are written in terms of orientation angles s and t of rays. We introduce a new function $K_{\pm}^*(x, y)$, $y \geq 0$, whose arguments are rectangular x and y coordinates of points on $\mathcal{C}$. Their meaning is as before. $\lambda K_{\pm}^*(x, y)$ is the expected number of disks centered between the rays $\mathbb{R}_x$ and $\mathbb{R}_{x \pm y}$, $y \geq 0$, which do not intersect $\mathbb{R}_x$. Explicit formula of $K_{\pm}^*(x, y)$, for the standard-uniform case, are given in Section 6.2. In the following section we derive the distribution of the length of a visible segment, he right of a visible point $\mathbf{P}_x$ on $\mathcal{C}$. In Section 6.3 we develop the formula of the distribution of the right-hand limit of a shadow on $\bar{\mathcal{C}}$, which is cast by a single random disk in C^*. In Section 6.4 we use the results of Section 6.3 and the distribution of residual length of shadows to obtain the distribution of the right-hand limit of a shadow starting at $\mathbf{P}_x$. These results are later used to obtain survival distributions of targets moving along linear paths. We discuss also the problem of determining the distribution of the number of visible and invisible segments on $\bar{\mathcal{C}}$.

6.1. The Distribution of The Length of A Visible Segment

Suppose that a point $\mathbf{P}_x$ on $\bar{\mathcal{C}}$ is visible. Let L_x denote the length of the visible segment of $\bar{\mathcal{C}}$, immediately to the right of $\mathbf{P}_x$. Let $V(l \mid x)$ denote the conditional c.d.f. of L_x, given that P_x is visible, i.e.,

$$V(l \mid x) = \Pr\{L_x \leq l \mid I(x) = 1\},$$

Let $\mu^* = \mu\{C^*\}$, be the expected number of disks centered at C^*. In order to find $V(l \mid x)$ we have to introduce the following definitions:

Let $\mathbf{C}_-(x)$ be the subset of C^* bounded by $\mathcal{U}$, $\mathcal{W}$ and the rays $\mathbb{R}_{x_L^*}$ and $\mathbb{R}_x$. Let $C(x,l)$ be the subset of C^* bounded by $\mathcal{U}$, $\mathcal{W}$ and the rays $\mathbb{R}_x$ and $\mathbb{R}_{x+l}$. Finally, let $C_+(x+l)$ be the subset of C^*, bounded by $\mathcal{U}$, $\mathcal{W}$ and $\mathbb{R}_{x+l}$, $\mathbb{R}_{x_U^*}$. We consider here values of l such that $0 < l < x_U^* - x$. Accordingly, for $0 \le l < x_U^* - x$,

$$\begin{aligned}\Pr\{L_x \ge l, I(x) = 1\} = \exp\{&-[\mu\{C_-(x)\} - \mu^* K_-^*(x, x - x_L^*)] \\ &- \mu\{C(x,l)\} - [\mu\{C_+(x+l) - \mu^* K_+^*(l+x, x_U^* - l - x)]\}.\end{aligned} \tag{6.1}$$

Finally, since $\mu^* = \mu\{C_-(x)\} + \mu\{C(x,l)\} + \mu\{C_+(x+l)\}$, and since

$$\psi(x) = \Pr\{I(x) = 1\} = \exp\{-\mu^*[1 - K_-^*(x, x - x_L^*) - K_+^*(x, x_U^* - x)]\}.$$

We obtain that, for $0 \le l < x_U - x$,

$$\begin{aligned}\bar{V}(l \mid x) &= 1 - V(l \mid x) \\ &= \frac{1}{\psi(x)} \Pr\{L_x \ge l, I(x) = 1\} \\ &= \exp\{-\mu^*[K_+^*(x, x_U^* - x) - K_+^*(x+l, x_U^* - x - l)]\}.\end{aligned} \tag{6.2}$$

Notice that $\bar{V}(l \mid x) = \bar{V}(l+x \mid 0)/\bar{V}(x \mid 0)$ for all $x_L^* \le x < l+x \le x_U^*$. In the following example we illustrate the function $\bar{V}(l \mid x)$ numerically.

Example 6.1. Consider the trapezoidal region with parameters $r = 100$ [m], $u = 50$ [m], $w = 75$ [m], $a = 1$ [m], $b = 2$ [m], $\lambda = 0.002$ [1/m^2] and $x_L = -100$ [m], $x_U = 100$ [m]. In the following table we present the values of $\bar{V}(l \mid x)$ for several values of x.

Table 6.1. Values of $\bar{V}(l \mid x)$ for Several x [m], l [m] values

l	$x = 0$	$x = 20$	$x = 40$
0	1.0000	1.0000	1.0000
5	0.8553	0.8546	0.8541
10	0.7313	0.7303	0.7294
15	0.6253	0.6239	0.6228
20	0.5345	0.5330	0.5317
25	0.4568	0.4552	0.4539
30	0.3903	0.3887	0.3874
35	0.3334	0.3319	0.3306
40	0.2849	0.2834	0.2822
45	0.2433	0.2419	0.2408
50	0.2078	0.2065	0.2054
55	0.1774	0.1762	0.1753
60	0.1515	0.1504	0.1663
65	0.1293	0.1283	—
70	0.1104	0.1095	—

Table 6.1. Values of $\bar{V}(l \mid x)$ for Several x [m], l [m] values (Continued)

l	$x = 0$	$x = 20$	$x = 40$
75	0.0942	0.0934	—
80	0.0804	0.0886	—
85	0.0686	—	—
90	0.0585	—	—
95	0.0499	—	—
100	0.0474	—	—

The values in Table 6.1 were computed with program VIEWLNG.

We see in Table 6.1 that the dependence of $V(l \mid x)$ on x is weak. In annular regions $V(l) = V(l \mid x)$ is independent of x. Also, the median length of the visibility segment under those conditions, is approximately 22 [m]. ■

6.2. The Functions $K^*_{\pm}(x,t)$ in the Standard-Uniform Case

Let $K_+(x,t,y)$ denote the area of the set bounded by the line $\mathcal{L}^+_x(y)$, the ray $\mathbb{R}_{x+t}$, $t \geq 0$, and the lines $\mathcal{U}$ and $\mathcal{W}$; $\mathcal{L}^+_x(y)$ is the line parallel to $\mathbb{R}_x$, on its r.h.s., of distance y from it. This is the set of all disk centers between $\mathbb{R}_x$ and $\mathbb{R}_{x+t}$, of radius $Y = y$, which do not intersect $\mathbb{R}_x$. In order to simplify notation, we assume that $w = r$. In actual computations we substitute xw/r and tw/r for x and t in the formulae given below. Let $d = (x^2 + w^2)^{1/2}$. Simple geometrical considerations yield:

$$\begin{aligned} K_+(x,t,y) = {} & I\left\{\frac{yd}{t} < u\right\}\left[\frac{w^2 - u^2}{2w}t - 2\frac{yd}{w+u}\right] \\ & + I\left\{u \leq \frac{yd}{t} < w\right\}\left[\frac{1}{2tw}(tw - yd)^2\right] \end{aligned} \tag{6.3}$$

where $I\{A\}$ is the indicator set function, which assumes the value 1 if A is true, and the value 0 otherwise.

Notice that $K_+(x,t,y)$ depends on x only via x^2. Symmetry implies that $K_-(-x,t,y) = K_+(x,t,y) = K_+(-x,t,y)$ for all $-\infty < x < \infty$. Hence, $K^*_+(x,t) = K^*_-(x,t)$ and we delete the $\pm$ subscript of K. Finally, $K^*(x,t) = \dfrac{1}{A\{C^*\}}E\{K(x,t,Y)\}$, where the expectation is with respect to the uniform distribution of Y over (a,b). $A\{C^*\}$ denotes the area of C^*. Let $x_1 = tu/d$ and $x_2 = tw/d$. The function $K^*(x,t)$ assumes the following forms:

(i) If $b < x_1$,

$$K^*(x,t) = \frac{w^2 - u^2}{2w}\left(t - \frac{d}{u+w}(a+b)\right) \div A\{C^*\}$$

(ii) If $a < x_1 < b \leq x_2$,

$$K^*(x,t) = \left\{ \frac{w^2-u^2}{2w}\left(t \cdot \frac{x_1-a}{b-a} - \frac{d}{u+w} \cdot \frac{1}{b-a}(x_1^2-a^2) \right) + \frac{1}{2tw}\left(t^2w^2\frac{b-x_1}{b-a} - tw\frac{d}{b-a}(b^2-x_1^2) + \frac{d^2}{3(b-a)}(b^3-x_1^3) \right) \right\} \div A\{C^*\} \tag{6.4}$$

(iii) If $a < x_1 < x_2 \leq b$,

$$K^*(x,t) = \left\{ \frac{w^2-u^2}{2w}\left(t\frac{x_1-a}{b-a} - \frac{d}{u+w}\frac{1}{b-a}(x_1^2-a^2) \right) + \frac{1}{2tw}\left(t^2w^2\frac{x_2-x_1}{b-a} - tw\frac{d}{b-a}(x_2^2-x_1^2) + \frac{d^2}{3(b-a)}(x_2^3-x_1^3) \right) \right\} \div A\{C^*\}$$

(iv) If $x_1 \leq a < b \leq x_2$,

$$K^*(x,t) = \left[\frac{tw}{2} - \frac{d}{2}(a+b) + \frac{d^2(a^2+ab+b^2)}{6tw} \right] \div A\{C^*\}$$

(v) If $x_1 \leq a < x_2 \leq b$,

$$K^*(x,t) = \left\{ \frac{tw}{2}\frac{x_2-a}{b-a} - \frac{d}{2(b-a)}(x_2^2-a^2) + \frac{d^2}{6tw(b-a)}(x_2^3-a^3) \right\} \div A\{C^*\}$$

(vi) If $x_2 < a$,

$$K^*(x,t) = 0.$$

6.3. Distribution of the Right-Hand Limit of a Shadow Cast By a Single Disk

Consider a shadow starting at a point $\mathbf{P}_x$ on $\bar{\mathcal{C}}$, which is cast by a single random disk We denote by $U_s(x)$ the coordinate of the right-hand limit of the shadow. A disk o radius Y, centered on a line $\mathcal{V}$, parallel to $\mathcal{U}$, of distance v from $\mathbf{O}$, $u \leq v \leq w$, casts a shadow on $\bar{\mathcal{C}}$, with left-hand limit at $\mathbf{P}_x$, if the x-coordinate of its center is (see Exercise [6.3.1])

$$x_c = \frac{x}{r}v + Y\left(1 + \left(\frac{x}{r}\right)^2\right)^{1/2} \tag{6.5}$$

The right-hand limit of this shadow is at coordinate

$$U_s(x) = r\ \tan\left(2\sin^{-1}\left(\frac{Y}{\sqrt{x_c^2+v^2}}\right) + \tan^{-1}\left(\frac{x}{r}\right)\right). \tag{6.6}$$

Substituting Eq. (6.5) in Eq. (6.6) and letting $Z = Y/V$, then the distribution of $U_s(x$ is like that of

$$S(Z;x) = r\ \tan\left(2\sin^{-1}\left(Z/\left(1+\left(\frac{x}{r}+Z\left(1+\left(\frac{x}{r}\right)^2\right)^{1/2}\right)^2\right)^{1/2}\right) + \tan^{-1}\left(\frac{x}{r}\right)\right) \tag{6.7}$$

where Z is distributed like Y/V. In the standard uniform case V and Y are independent, $Y \sim \mathcal{U}(a,b)$ and $V \sim \mathcal{U}(u,w)$. Notice that $U_m(x) \le U_s(x) \le U_M(x)$, where $U_m(x)$ is obtained by substituting $Z = \dfrac{a}{w}$ in Eq. (6.7) and $U_M(x)$ is obtained by substituting $Z = b/u$ in Eq. (6.5). Moreover, $S(z;x)$ is a strictly increasing function of z, as can be verified by differentiating $S(z;x)$ with respect to z. Accordingly, the distribution of $U_s(x)$ can be determined from the distribution of Z. Let $H_Z(z)$ denote the c.d.f. of Z. It is straightforward to prove (see Problem [6.3.2]) that

$$\begin{aligned} H_Z(z) = \frac{1}{(w-a)(b-a)} \Bigg[\frac{z}{2} \Bigg(\left(\min\left(w, \frac{b}{z}\right)\right)^2 \\ - \left(\max\left(u, \frac{a}{z}\right)\right)^2 \Bigg) - a \left(\min\left(w, \frac{b}{z}\right) - \max\left(u, \frac{a}{z}\right)\right)\Bigg] \\ + I\left\{ z > \frac{b}{w} \right\} \frac{w - b/z}{w-u}. \end{aligned} \tag{6.8}$$

Let $z(\eta;x)$ denote the inverse function of $S(z;x)$. Simple algebra yields

$$z(\eta;x) = \frac{\psi^2(\eta)\frac{x}{r}(1+(\frac{x}{r})^2)^{1/2} + \psi(\eta)((1+(\frac{x}{r})^2)(1-\psi^2(\eta))^{1/2}}{1-\psi^2(\eta)(1+(\frac{x}{r})^2)} \tag{6.9}$$

where

$$\psi(\eta) = \sin\left(\frac{1}{2}\left(\tan^{-1}\left(\frac{\eta}{r}\right) - \tan^{-1}\left(\frac{x}{r}\right)\right)\right), \quad x \le \eta \tag{6.10}$$

Let $Q_s(\eta \mid x) = P\{U_s(x) \le \eta\}$ denote the c.d.f. of $U_s(x)$, when the shadow is cast by a single random disk. Then,

$$Q_s(\eta \mid x) = H_Z(z(\eta;x)) \tag{6.11}$$

In Table 6.2 we present the c.d.f. $Q_s(\eta \mid x)$, for the case of a trapezoidal region with parameters $a = 1$ [m], $b = 2.5$ [m], $u = 40$ [m], $w = 60$ [m], $r = 100$ [m] and $x =$ [illegible] [m]. Program SNGSHDW computes this c.d.f. The output is printed also in file SNGSADW.DAT.

Table 6.2. The c.d.f. $Q_s(u \mid x)$.

u	$Q_s(u \mid s)$
8.7823	0.01535
9.2234	0.05497
9.6644	0.11195
10.1055	0.18156
10.5466	0.25464
10.9877	0.32767
11.4288	0.40065
11.8699	0.47359
12.3109	0.54649

Table 6.2. The c.d.f. $Q_s(u \mid x)$. (Continued)

u	$Q_s(u \mid s)$
12.7520	0.61933
13.1931	0.69211
13.6342	0.76256
14.0753	0.82161
14.5163	0.86983
14.9574	0.90865
15.3985	0.93929
15.8396	0.96275
16.2807	0.97988
16.7217	0.99140
17.1628	0.99793
17.6039	1.00000

6.4. Distribution of the Right-hand Limit of a Shadow Starting at a Given Point

The right-hand limit $U_s(x)$ may not be the right-hand point of the shadow, if the number of disks centered to the right of $\mathbb{R}_x$ is greater than one. Consider a particular disk whose shadow starts at $\mathbf{P}_x$ and ends at a point with coordinate $U_s(x) = \eta$, $x < \eta$. The shadow to the right of $\mathbf{P}_x$ ends at a point P_t, with $t > \eta$, if some shadows cast by other disks start to the left of $\mathbf{P}_\eta$ and terminate to the right of $\mathbf{P}_\eta$. We employ now concepts introduced by Chernoff and Daly (1957) to derive the distribution of right-hand point of the residual shadow, to the right of $\mathbf{P}_\eta$.

Consider the rays $\mathbb{R}_\eta$ and $\mathbb{R}_t$, with $t > \eta > x$. Let $N_+(x, \eta, t)$ denote the number of disks, centered to the right of $\mathbb{R}_x$, which intersect both $\mathbb{R}_\eta$ and $\mathbb{R}_t$. Define the functional

$$T(\eta) = \sup\{t : N_+(x, \eta, t) \geq 1\}. \qquad (6.12)$$

Let $T^0(\eta) = \eta$ and $T^{i+1}(\eta) = T(T^i(\eta))$. Obviously, $T^{i+1}(\eta) \geq T^i(\eta)$ for all $i = 0, 1, \cdots$. Let $U_R(\eta) = \lim_{i\to\infty} T^i(\eta)$. $U_R(\eta)$ is the coordinate of the right-hand limit of the residual shadow, to the right of $\mathbf{P}_\eta$. Clearly, $\{T(\eta) > t\} = \{N_+(x, \eta, t) \geq 1\}$. Let $\mu_+(x, \eta, t)$ be the expected value of $N_+(x, \eta, t)$. Then, from the Poisson field assumption, the c.d.f. of $T(\eta)$ is

$$H_1(t; x, \eta) = P_x\{T(\eta) \leq t\} = \exp\{-\mu_+(x, \eta, t)\}. \qquad (6.13)$$

Let $\mu_+^*(x) = \lambda(x_U^* - x)(w^2 - u^2)/2r$, denote the expected number of disks in C^*, to the right of $\mathbb{R}_x$. Since $\mathbf{P}_x$ is the left-hand limit of a shadow, we have to consider only disks centered to the right of $\mathbb{R}_x$, which do not intersect it. The expected number of such disks is $\mu_+^*(x)K^*(x, x_U^* - x)$. If we subtract from this quantity the expected number of disks which do <u>not</u> intersect either $\mathbb{R}_\eta$ or $\mathbb{R}_t$, we obtain

$$\begin{aligned} \mu_+(x, \eta, t) = \mu_+^*(x)[K^*(x, x_U^* - x) - K^*(\eta, x_U^* - \eta) \\ - K^*(t, t - \tilde{t}_x) - K^*(x, \tilde{t}_x - x) + K^*(\eta, \tilde{t}_\eta - \eta) \\ + K^*(t, t - \tilde{t}_\eta)], \end{aligned} \qquad (6.14)$$

where $\tilde{t}_x$ is the coordinate of the point intersecting the bi-sector between $\mathbb{R}_x$ and $\mathbb{R}_t$, and $\tilde{t}_\eta$ is that between $\mathbb{R}_\eta$ and $\mathbb{R}_t$, i.e.,

$$\tilde{t}_x = r \tan\left(\left(\tan^{-1}\left(\frac{t}{r}\right) + \tan^{-1}\left(\frac{x}{r}\right)\right)/2\right),$$

and

$$\tilde{t}_\eta = r \tan\left(\left(\tan^{-1}\left(\frac{t}{r}\right) + \tan^{-1}\left(\frac{\eta}{2}\right)\right)/2\right).$$

Notice that

$$\begin{aligned}\lim_{t\downarrow\eta} H_1(t;x,\eta) &= \exp\{-\mu_+^*(x)[K^*(x,x_U^* - x) \\ &- K^*(\eta, x_U^* - \eta) - K^*(\eta, \eta - \tilde{\eta}_x) - K^*(x, \tilde{\eta}_x - x)]\}.\end{aligned} \tag{6.15}$$

This is the probability that the ray $\mathbb{R}_\eta$ is not intersected by disks to the right of $\mathbb{R}_x$, given that $\mathbb{R}_x$ is not intersected by these disks. Thus, the c.d.f. of $T(\eta)$, $H_1(t;x,\eta)$ is zero for $t < \eta$, it has a jump $H_1(\eta;x,\eta)$, at $t = \eta$, and is absolutely continuous for $t > \eta$. This property is inherited by the c.d.f. of $T^i(\eta)$, $H_i(t;x,\eta)$ $i \geq 1$. We give now the recursive relationship between $H_i(t;x,\eta)$ and $H_{i-1}(t;x,\eta)$ $i \geq 2$.

Introduce the bivariate distribution

$$G_n(t_1,t_2;x,\eta) = P_x\{T^{n-1}(\eta) \leq t_1, T^n(\eta) \leq t_2\}.$$

Since $\{T^{(n)}(\eta) \leq t\} \subset \{T_{n-1}^{(n)}(\eta) \leq t^*\}$ for all $t^* \geq t$,

$$\begin{aligned}H_n(t;x,\eta) &= P_x\{T^n(\eta) \leq t\} \\ &= G_{n-1}(t^*,t;x,\eta)\end{aligned} \tag{6.16}$$

Moreover, for all $\eta < z < y < t$,

$$\begin{aligned}&P_x\{T^n(\eta) \leq t \mid T^{n-2}(\eta) = z, T^{n-1}(\eta) = y\} \\ &\qquad = \exp\{-[\mu_+(x,y,t) - \mu_+(x,z,t)]\}.\end{aligned} \tag{6.17}$$

Indeed, given that $\{T^{n-2}(\eta) = z$ and $T^{n-1}(\eta) = y\}$, $\{T^n(\eta) > t\}$ if, and only if, there exists at least one disk which intersects $\mathbb{R}_y$ and $\mathbb{R}_t$ and does not intersect $\mathbb{R}_z$. Accordingly,

$$\begin{aligned}G_n(t_1,t_2;x,\eta) = \int_\eta^{t_1}\int_z^{t_1} &\exp\{-[\mu_+(x,u,t_2) - \mu_+(x,z,t_2)]\} \\ &\cdot dG_{n-1}(z,u;x,\eta).\end{aligned} \tag{6.18}$$

Notice that

$$G_1(t_1,t_2;x,\eta) = \begin{cases} H_1(t_2;x,\eta), & \text{if } t_1 = \eta \\ \\ 0, & \text{if } t_1 \neq \eta \end{cases} \tag{6.19}$$

Accordingly,

$$G_2(t_1,t_2;x,\eta) = \exp\{\mu_+(x,\eta,t_z) \cdot \int_\eta^{t_1} \exp\{-\mu_+(x,u,t_2)\}dH_1(u;\eta,x), \tag{6.20}$$

and

$$H_2(t;x,\eta) = G_2(t,t;x,\eta).$$

Finally, for each t, $H_n(t;x,\eta) \leq H_{n-1}(t;x,\eta)$ for all $n \geq 2$. The limit of $H_n(t;x,\eta)$ is the required c.d.f.

$$D_R(t;x,\eta) = \lim_{n\to\infty} H_n(t;x,\eta).$$

A shadow starting at $\mathbf{P}_x$ terminates at $\mathbf{P}_{U_T(x)}$ where $U_T(x) = U_R(U_s(x))$. Thus, if we denote by $D_T(t \mid x)$ the c.d.f. of $U_T(x)$, we obtain

$$D_T(t \mid x) = \int_{U_m(x)}^{\min(t,U_M(x))} D_R(t;x,\eta)dQ_s(\eta \mid x). \tag{6.21}$$

6.5. Discrete Approximation

The distribution functions $D_R(t;x,\eta)$ and $D_T(t \mid x)$ are approximated by the following discrete numerical scheme.

For a given coordinate x of the starting point (left-hand limit) of a shadow, we compute first the minimal and maximal values of $U_s(x)$, i.e. $U_m(x)$ and $U_M(x)$. The interval $(U_m(x), U_M(x))$ is partitioned into N subintervals of equal size.

Let $\delta = (U_M(x) - U_m(x))/N$, and for $i = 0, \cdots, N$, let $u_i = U_m(x) + i\delta$. Since the total shadow ends to the right of $U_M(x)$, introduce another integer $N^* > N$, and let $t_i = U_m(x)+i\delta$, $i = 0, \cdots, N^*$. The c.d.f. $D_T(t \mid x)$ given by Eq. (6.21) is approximated by

$$\hat{D}_T(t_i \mid x) = \sum_{j=1}^{\min(i,N)} p_s(j)\hat{D}_R(t_i;x,U_m(x) - \frac{\delta}{2} + j\delta), \tag{6.22}$$

where, for $j = 1, \cdots, N$,

$$p_s(j) = Q_s(u_j \mid x) - Q_s(u_{j-1} \mid x), \tag{6.23}$$

and $\hat{D}_R(t_i;x,\eta_j)$ is a discrete approximation to $D_R(t;x,\eta)$, $\eta_j = U_m(x) - \frac{\delta}{2} + j\delta$. The function $\hat{D}_R(t_i;x,\eta_j)$ is determined in the following manner.

In the first stage compute $H_1(t \mid x,\eta)$ on a discrete lattice. Thus, for $j = 1, \cdots, N$ and $i = j, \cdots, N^*$, let

$$\hat{H}_1(i,j) = \exp(-\mu_+(x,u_j - \frac{\delta}{2},t_i)).$$

In the next stage we approximate $H_2(t \mid x, \eta)$ by

$$
\begin{aligned}
\hat{H}_2(i,j) = \sum_{k=j}^{i} \exp\{-\mu_+(x, u_k - \frac{\delta}{2}, t_i) \\
+ \mu_+(x, u_j - \frac{\delta}{2}, t_i)\}(\hat{H}_1(k,j) - \hat{H}_1(k-1,j)),
\end{aligned}
\tag{6.24}
$$

$j = 1, \cdots, N$, $i = j, \cdots, N^*$.

In order to approximate the functions $H_n(t; x, \eta)$ for $n \geq 3$ we need to develop a more complicated scheme. We approximate the functions $G_n(t_1, t_2; x, \eta)$ by computing recursively,

$$
\begin{aligned}
\hat{G}_n(i_1, i_2, j) = \sum_{k=j}^{i_1} \sum_{l=k}^{i_1} \exp\{-[\mu_+(x, u_l - \frac{\delta}{2}, t_{i_2}) \\
- \mu_+(x, u_k - \frac{\delta}{2}, t_{i_2})]\}[\hat{G}_{n-1}(k, l, j) \\
- \hat{G}_{n-1}(k-1, l, j) - \hat{G}_{n-1}(k, l-1, j) + \hat{G}_{n-1}(k-1, l-1, j)],
\end{aligned}
\tag{6.25}
$$

$i_1 = j, \cdots, N^*$, $i_2 = i_1, \cdots, N^*$, $j = 1, \cdots, N$, where

$$
\begin{aligned}
\hat{G}_2(i_1, i_2, j) = \sum_{k=j}^{i_1} \exp\{-[\mu_+(x, u_k - \frac{\delta}{2}, t_{i_2}) \\
- \mu_+(x, u_j - \frac{\delta}{2}, t_{i_2})]\}(\hat{H}_1(k,j) - \hat{H}_1(k-1,j)).
\end{aligned}
\tag{6.26}
$$

Finally we obtain

$$
\hat{H}_n(i,j) = \hat{G}_n(i,i,j), \quad i = j, j+1, \cdots, N^*. \tag{6.27}
$$

In Table 6.3 we present the values of $\hat{H}_n(i,j)$, $n = 1, \cdots, 6$, for a fixed value of j. The calculations were done with the following field parameters in [m]: $x_L^* = -100$, $x_U^* = 300$, $c = 10$, $\eta = 15$, $r = 100$, $u = 40$, $w = 60$, $a = 2$, $b = 3.5$ and $\lambda = 0.01$ [1/m^2]. In these calculations $\delta = 2$ [m].

Table 6.3. The Functions $\hat{H}_n(i,j)$

i	$n=1$	$n=2$	$n=3$	$n=4$	$n=5$	$n=6$
0	.6087	.6087	.6087	.6087	.6087	.6087
2	.6337	.6300	.6300	.6292	.6292	.6292
4	.6862	.6667	.6667	.6643	.6643	.6643
6	.7577	.7097	.7094	.7050	.7050	.7050
8	.8361	.7550	.7531	.7468	.7468	.7468
10	.9105	.8014	.7957	.7875	.7875	.7875
12	.9648	.8456	.8329	.8232	.8232	.8232
14	.9908	.8855	.8625	.8516	.8516	.8516

Table 6.3. The Functions $\hat{H}_n(i,j)$ (Continued)

i	$n=1$	$n=2$	$n=3$	$n=4$	$n=5$	$n=6$
16	.9999	.9205	.8865	.8743	.8743	.8743
18	1	.9493	.9071	.8931	.8931	.8931
20	1	.9709	.9257	.9092	.9090	.9090
22	1	.9853	.9427	.9230	.9226	.9226
24	1	.9936	.9578	.9349	.9340	.9340
26	1	.9977	.9707	.9454	.9437	.9436
28	1	.9993	.9809	.9547	.9518	.9517
30	1	.9999	.9884	.9600	.9580	.9578

Notice that the value i in Table 6.3 corresponds to the length of the residual shadow [m], i.e., $U_R(15) - 15$. We see in the table that after $n = 4$ iterations the functions change very slightly. It seems that four iterations are sufficient for the computation of $\hat{D}_R(t \mid x, \eta)$. We therefore approximate Eq. (6.22) by

$$\hat{D}_T(t_i \mid x) = \sum_{j=1}^{\min(i,N)} p_s(j)\hat{H}_4(i,j) \tag{6.28}$$

The amount of computations required to determine $\hat{D}_T(t_i \mid x)$ according to (6.28) is large. We have found that the following approximation works well, and reduces the length of computations, i.e., replace $\hat{H}_4(i,j)$ by $\tilde{H}_4(i,j)$ where

$$\tilde{H}_4(0,j) = \hat{H}_2(0,j),$$

and for $i = 1, 2, \cdots$

$$\tilde{H}_4(i,j) = \tilde{H}_4(i-1,j) + .88(\hat{H}_2(i,j) - \hat{H}_2(i-1,j)).$$

In Table 6.4 we present the c.d.f. $D_T(t \mid x)$ for the case of $r = 100$ [m], $u = 40$ [m], $w = 60$ [m], $x_L^* = -100$ [m], $x_U^* = 300$ [m], $a = 2$ [m], $b = 3.5$ [m], $x = 10$ [m] and 3 values of λ.

Table 6.4. The c.d.f. of the right-hand limit of a shadow starting at $x = 10$ [m].

	λ		
t	0.001	0.005	0.01
15.00	0.00000	0.00000	0.00000
20.00	0.22322	0.16945	0.11986
25.00	0.87452	0.66056	0.45645
30.00	0.97437	0.83600	0.64166
35.00	0.99352	0.92195	0.78269

Table 6.4. The c.d.f. of the right-hand limit of a shadow starting at $x = 10$ [m]. (Continued)

	λ		
40.00	0.99858	0.97195	0.90886
45.00	0.99973	0.99384	0.97808
50.00	0.99997	0.99932	0.99742
55.00	1.00000	0.99997	0.99989
60.00	1.00000	1.00000	1.00000

The c.d.f. $\hat{D}_T(t \mid x)$ was computed with program CDFSHDW. From this table we can obtain the numerical values of the expected value and standard deviation of the shadow length. These are given in the following table.

λ	0.001	0.005	0.01
Expected Value	12.18	14.73	18.08
Std. Dev.	3.42	6.02	7.82

6.6. Distribution of the Number of Shadows

Consider the trapezoidal region, and let $\mathbf{P}_x$ and $\mathbf{P}_y$ be two points in $\bar{\mathcal{C}}$, $x_L < x < y < x_U$. Let $J(x, y)$ denote the number of shadows on the line segment between $\mathbf{P}_x$ and $\mathbf{P}_y$. We will assume here that the point $\mathbf{P}_x$ is visible and will develop the conditional probability that $J(x, y) = j$, given that $\{I(x) = 1\}$. Let $P_j(x, y)$ denote this conditional probability. Obviously

$$P_0(x, y) = \bar{V}(y - x \mid x). \tag{6.29}$$

visible segment is followed to the right by a shadowed segment. Suppose that the isible segment to the right of $\mathbf{P}_x$ ends at $\mathbf{P}_t$. The right-hand limit of the following adow is at $U_T(t)$.

Let $W(y \mid x)$, $y \geq x$, denote the c.d.f. of the right-end limit of a cycle of visible gment followed by a shadowed segment.

$$\begin{aligned} W(y \mid x) &= \int_x^y D_T(y \mid t) dV(t - x \mid x) \\ &= \int_0^{y-x} D_T(y \mid u + x) dV(u \mid x), \end{aligned} \tag{6.30}$$

here the distribution function $V(y \mid x)$ and $D_T(y \mid x)$ are given by Eqs. (6.2) and 21). The conditional probability that $J(x, y) = 1$, given that $\{I(x) = 1\}$ is

$$\begin{aligned} P_1(x, y) = &\int_x^y \bar{V}(y - t \mid t) dW(t \mid x) \\ &+ \int_x^h (1 - D_T(y \mid t)) dV(t - x \mid x), \end{aligned} \tag{6.31}$$

Finally, for each $j \geq 2$ we can determine $P_j(x, y)$ recursively, by the equation

$$P_j(x, y) = \int_x^y P_{j-1}(t, y) dW(t \mid x). \tag{6.32}$$

The c.d.f. $W(y \mid x)$ and the probabilities $P_j(x, y)$, for $j = 0, 1, 2$ can be computed by program DISTNSH. In Table 6.5 we present a discrete approximation to $W(y \mid x)$ for the parameters $r = 100$ [m], $u = 40$ [m], $w = 60$ [m], $a = 2$ [m], $b = 3.5$ [m], $\lambda = 0.01$ [1/m^2] and $x = 30$ [m].

Table 6.5. Values of the Cycle C.D.F. $W(y \mid x)$

y	$W(y \mid x)$	y	$W(y \mid x)$
4	0.0000	36	0.6511
8	0.0000	40	0.7379
12	0.0211	44	0.8099
16	0.1043	48	0.8645
20	0.2219	52	0.9056
24	0.3222	56	0.9329
28	0.4321	60	0.9520
32	0.5420	—	—

For these field parameters we find $P_0(30, 90) = 0.002$, $P_1(30, 90) = 0.135$, $P_2 = 0.595$. Thus, the probability is 0.268 of having three or more shadows between 30 and 90 [m]

6.7. Survival Probability Functions

Consider a target moving along $\mathcal{C}$, from the point $\mathbf{P}_x$ to the point $\mathbf{P}_y$, $x^* < x < y < x^{**}$ Suppose that $\mathbf{P}_x$ is visible. A hunter, standing at the origin starts shooting at the target when the target is at $\mathbf{P}_x$. The hunter can continue shooting as long as the target is within a visible segment of $\mathcal{C}$. Let e^{-qL}, $0 < q < \infty$, denote the probability that the target, moving through a visible segment of length L, will survive (not be hit). Let $S(x, y)$ designate the survival probability function.

We distinguish between the exclusive and exhaustive events:

(i) The visible segments to the right of $\mathbf{P}_x$ terminates to the right of $\mathbf{P}_y$;

(ii) The visible segments to the right of $\mathbf{P}_x$ terminates at $\mathbf{P}_t$, $t < y$, and the length of the shadow starting at $\mathbf{P}_t$ is larger than $y - t$;

(iii) The visible segment to the right of $\mathbf{P}_x$ terminates at $\mathbf{P}_t$, $t < y$, and the length of the shadow starting at $\mathbf{P}_t$ is smaller than $y - t$.

The survival probability is given, accordingly by the equation

$$\begin{aligned} S(x, y) &= e^{-q(y-x)} \bar{V}(y - x \mid x) \\ &+ \int_x^y e^{-q(t-x)} (1 - D_T(y \mid t)) dV(t - x \mid x) \\ &+ \int_x^y e^{-q(t-x)} \left\{ \int_{U_m(t)}^y S(z, y) dD_T(z \mid t) \right\} dV(t - x \mid x), \end{aligned} \tag{6.33}$$

where $\bar{V}(\cdot, x)$ given by Eq. (6.2), and $D_T(\cdot \mid x)$ is given by Eq. (6.21).

Let $t_m(x)$ be the inverse function of $U_m(x)$. Define the functions

$$A(x,y) = e^{-q(y-x)}\bar{V}(y-x \mid x) + \int_0^{y-x} e^{-qt}(1 - D_T(y \mid t+x))dV(t \mid x), \tag{6.34}$$

and for $z \geq U_m(x)$,

$$B(x,z) = \int_0^{t_m(z)-x} e^{-qt}\left(\frac{\partial}{\partial z}D_T(z \mid t+x)\right)dV(t \mid x), \tag{6.35}$$

then, equation (6.33) can be written as

$$S(x,y) = A(x,y) + \int_{U_m(x)}^{y} S(z,y)B(x,z)dz. \tag{6.36}$$

The survival function $S(x,y)$ can now be approximated by a discrete algorithm, similar to that of the previous section.

For a given integer N, let $\delta = (y-x)/N$, and let $x_i = x + i\delta$, $(i = 0, \cdots, N)$. Define recursively,

$$\hat{A}(N,N) = 1,$$

$$\hat{A}(N-1,N) = e^{-q\delta}\bar{V}(\delta \mid x_{N-1}) + e^{-q\delta/2}(1 - \hat{D}_T(x_N \mid x_N - \frac{\delta}{2}))V(\delta \mid x_{N-1}),$$

and for $j = 2, \cdots, N$

$$\begin{aligned}\hat{A}(N-j,N) = {} & e^{-j\delta q}\bar{V}(j\delta \mid x_{N-j}) \\ & + \sum_{i=1}^{j} e^{-q(l-\frac{1}{2})\delta}[V(l\delta \mid x_{N-j}) \\ & - V((l-1)\delta \mid x_{N-j})](1 - \hat{D}_T(x_N \mid x_{N-j+l} - \frac{\delta}{2}))\end{aligned} \tag{6.37}$$

In addition, let

$$\hat{B}(N,N) = 0,$$

$$\hat{B}(N-1,N) = e^{-q\delta/2}V(\delta \mid x_{N-1})\hat{D}(x_N \mid x_N - \frac{\delta}{2}),$$

and for $j = 2, \cdots, N$, $l = 1, \cdots, j$,

$$\hat{B}(N-j, N-j+l) = \sum_{i=1}^{l} e^{-q\delta(i-\frac{1}{2})}[V(i\delta \mid x_{N-j}) - V((i-1)\delta \mid x_{N-j})][\hat{D}(x_{N-j+l} \mid x_{N-j+i} - \frac{\delta}{2}) - I\{i < l\}\hat{D}(x_{N-j+l-1} \mid x_{N-j+i} - \frac{\delta}{2})]. \tag{6.38}$$

Then, the survival function can be approximated recursively by

$$\begin{aligned} \hat{S}(N,N) &= 1, \\ \hat{S}(N-j,N) &= \hat{A}(N-j,N) + \hat{B}(N-j,N) \\ &+ I\{j \geq 2\} \sum_{i=N-j+1}^{N-1} \hat{B}(N-j,i)\hat{S}(i,N). \end{aligned} \tag{6.39}$$

In Table 6.5 we present the survival probabilities, computed according to Eqs. (6.37) - (6.39) for the field parameters $r = 100$ [m] $u = 40$ [m], $w = 60$ [m], $a = 2$ [m], $b = 3.5$ [m], $t_x = 0$, and various values of λ. As before, $x_L = -100$ [m] and $x_U = 100$ [m]. The values of this table were computed with program SURVFUC.

Table 6.6. Survival Probabilities, $\hat{S}(N-j, N)$, for $\lambda = 0.01$ (0.01) 0.05, with $e^{-q} = 0.8$

$j \backslash \lambda$	0.01	0.02	0.03	0.04
0	1.0000	1.0000	1.0000	1.0000
1	0.8091	0.8173	0.8247	0.8314
2	0.6710	0.6978	0.7211	0.7413
3	0.5711	0.6197	0.6599	0.6932
4	0.4989	0.5686	0.6236	0.6674
5	0.4466	0.5352	0.6021	0.6536
6	0.4088	0.5133	0.5894	0.6462
7	0.3814	0.4989	0.5818	0.6421
8	0.3515	0.4801	0.5708	0.6359
9	0.3399	0.4771	0.5706	0.6365
10	0.3295	0.4733	0.5693	0.6360
11	0.3056	0.4584	0.5615	0.6322
12	0.3106	0.4582	0.5620	0.6326
13	0.2941	0.4542	0.5599	0.6316
14	0.2790	0.4455	0.5559	0.6299
15	0.2711	0.4406	0.5533	0.6286
16	0.2636	0.4369	0.5519	0.6281
17	0.2513	0.4286	0.5474	0.6259
18	0.2434	0.4245	0.5458	0.6253
19	0.2331	0.4170	0.5414	0.6229
20	0.2247	0.4122	0.5393	0.6220

Notice that if there is complete visibility then the survival probabilities for a target moving $L = 2$ [m] is 0.64. We see in Table 6.4 that when λ changes from 0.01 [1/m^2] to 0.05 [1/m^2] then the survival probabilities for $L = 2$ [m] increase from 0.67 to 0.74. Moreover, if $\lambda = 0.01$ the probability that a target will survive, when $L = 18$ [m] is only 0.19 while when $\lambda = 0.05$ it increases to 0.64. We assumed here that the target is detected immediately when it enters a visible segment. In actual applications one has to allow some time for detection, which increases the survival probabilities.

7
PROBLEMS AND SOLUTIONS

In the present chapter we present problems for solution for each one of the previous six chapters. The problems are followed by the solution of the problem. Often the solutions are detailed. This is done in order to further explain the methodology. The furnished software is being utilized extensively in solving the problems. In solving problems of Chapter 2 we often use numerical integration. For this we have used either MATHCAD 4.0® or MATHEMATICA®.

The problems are numbered by the chapter section and sequential number, e.g. problem [1.1.1] is the first problem for Chapter 1, Section 1.

7.1.1. Problems for Chapter 1

[1.1.1] Consider a square S with side length a. Let C be a circle centered at the intersection of the two diagonals of S, and passing through the four vertices of S. (C inscribes S.) A point is chosen in C at random, so that the probability that the point belongs to a subset B of C, is proportional to the area of B. What is the probability that the point does *not* belong to S?

[1.1.2] A stick of length l is randomly broken at two points. What is the probabilit that a triangle can be constructed from the three pieces of the broken stick?

[1.1.3] Two random variables X, Y have a uniform joint distribution over the regio $S = \{-1 \le x \le 1, -1 \le y \le 1\}$, i.e.,

$$f(x,y) = \begin{cases} \frac{1}{4}, & \text{if } (x,y) \in S \\ 0, & \text{otherwise.} \end{cases}$$

Find:
(i) $\Pr\{X + Y \le 1\}$
(ii) $\Pr\{X^2 + Y^2 \le 1\}$
(iii) $\Pr\{-1 \le X + Y \le 1 \mid X^2 + Y^2 \le 1\}$.

[1.1.4] Disks are randomly scattered over a triangular region T. The vertices of are $(-1,0)$, $(1,0)$ and $(0,1)$. The centers of the disks (X,Y) are uniform distributed over T. The radii of disks have a distribution $G(r \mid x)$, whi depends on the x-coordinate of their center, with p.d.f.

$$g(r \mid x) = \frac{30r^2(1-r)^2}{10(1-|x|)^3 - 15(1-|x|)^4 + 6(1-|x|)^5}$$

for $0 < r < 1 - |x|$; $-1 < x < 1$. What is the probability that a random di will not intersect the line connecting the origin $(0,0)$ with the vertex $(0,1)$?

[1.2.1] Let X have a Binomial distribution with parameters $N = 45$ and $\theta = .17$. U program DISTRIB to find
(i) $\Pr\{4 \le X \le 10\}$

(ii) $\Pr\{X > 13\}$

(iii) Sum the values of $\Pr\{X > a\}$ for $a = 1, 2, \cdots, N$. Is this sum close to the value of $N\theta$?

[1.2.2] Let X have a Poisson distribution with $\lambda = 10$. Use program DISTRIB to determine

(i) $\Pr\{X > 10\}$

(ii) The quartiles of X

(iii) Is the sum of $\Pr\{X \geq a\}$, for $a = 0, 1, 2, \cdots$, close to the value of λ?

[1.2.3] A certain region $\mathcal{S}$ is partitioned into three subregions $\mathcal{S}_1$, $\mathcal{S}_2$, $\mathcal{S}_3$. The numbers of obscuring objects in these regions are Poisson random variables with means $\lambda_1 = 10$, $\lambda_2 = 15$, $\lambda_3 = 20$, respectively. Assuming that these random variables are independent, find the probability that there are at least 50 obscuring objects in $\mathcal{S}$.

[1.2.4] The numbers of trees, N, in a given section of a forest, is a random variable having a Poisson distribution with mean λ. The diameters of the trunks of trees, .1[m] above the ground, are independent random variables having a distribution with c.d.f. $F(x)$ on the interval $[a, b]$. What is the distribution of the j-th smallest diameter?

[1.2.5] In continuation of Problem [1.2.4] show that, if the radii of trees are uniformly distributed over $[0, b]$ then the conditional expected value of the j-th smallest radius, given $\{N \geq j\}$, is

$$E\{X_{(j)} \mid N \geq j\} = \frac{b}{\lambda} \cdot \frac{j(1 - P(j; \lambda))}{1 - P(j-1; \lambda)}.$$

[1.2.6] Show that the c.d.f. of the Binomial distribution satisfies

$$1 - B(k; n, \theta) = I_\theta(k+1, n-k), \qquad k = 0, 1, \cdots, n-1.$$

[Hint: Apply integration by parts on the incomplete beta integral.]

[1.2.7] Establish by integration by parts that

$$P\{G(1, k+1) \leq x\} = 1 - P(k; x),$$

where $G(1, k+1)$ is a Gamma random variable with $\lambda = 1$, $\nu = k+1$; and $P(k; x)$ is the c.d.f. of a Poisson distribution with mean x.

[1.2.8] (i) Show that the moment generating function of a Normal distribution $N(\mu, o)$ is

$$M(t) = E\{e^{tX}\} = \exp\left\{t\mu + \frac{t^2\sigma^2}{2}\right\}.$$

1.2.9] Show that if X is distributed like $N(0,1)$ then X^2 is distributed like $\chi^2[1]$.

.2.10] Let (X, Y) have a bivariate normal distribution with parameters $(\xi, \eta, \sigma_x, \sigma_y, \rho)$. Show that the conditional distribution of X, given $\{Y = y\}$, is normal with mean

$$E\{X \mid Y = y\} = \xi + \rho\frac{\sigma_x}{\sigma_y}(y - \eta),$$

and variance

$$V\{X \mid Y = y\} = \sigma_x^2(1 - \rho^2).$$

Establish Eq. (1.46).

[1.2.11] Let σ_x^2, σ_y^2, ρ be the dispersion parameters (variances and correlation) in a bivariate normal distribution. Let $\sigma_{xy} = \rho\sigma_x\sigma_y$ be the covariance. The matrix

$$V = \begin{bmatrix} \sigma_x^2 & \sigma_{xy} \\ \sigma_{xy} & \sigma_y^2 \end{bmatrix}$$

is called the **covariance matrix**. Show that the bivariate normal p.d.f. can be presented, in matrix notation, by the formula

$$f(x,y) = \frac{1}{(2\pi)|V|^{1/2}} \exp\left\{-\frac{1}{2}(x-\xi, y-\eta)V^{-1}\begin{pmatrix} x-\xi \\ y-\eta \end{pmatrix}\right\},$$

where $|V| = \sigma_x^2\sigma_y^2(1-\rho^2)$, is the determinant of V.

[1.2.12] The moment generating function of a bivariate distribution is defined as

$$M(t_1,t_2) = E\{\exp\{t_1X + t_2Y\}\}.$$

Prove that the moment generating function of a bivariate normal distribution is

$$M(t_1,t_2) = \exp\{t_1\xi + t_2\eta + \frac{1}{2}(t_1^2\sigma_x^2 + t_2^2\sigma_y^2 + 2t_1t_2\rho\sigma_x\sigma_y)\}.$$

[1.2.13] Let $\mathbf{X} = \begin{pmatrix} X_1 \\ X_2 \end{pmatrix}$ and $\mathbf{U} = A\mathbf{X}$, where A is a 2×2 non-singular matrix. Assume that $\mathbf{X}$ has a bivariate normal distribution with mean vector $\boldsymbol{\xi} = \begin{pmatrix} \xi_1 \\ \xi_2 \end{pmatrix}$, i.e., $\xi_i = E\{X_i\}$, $i = 1,2$, and covariance matrix V. Prove that $\mathbf{U}$ has a bivariate normal distribution, with mean vector $A\boldsymbol{\xi}$ and covariance matrix AVA', where A' is the transpose of A.

[1.2.14] Show that in a bivariate normal distribution, (X,Y) are independent if, and only if, the covariance matrix V is diagonal (i.e., $\rho = 0$).

[1.2.15] Two points (x_1,y_1) and (x_2,y_2) are independent and distributed according to the bivariate normal distribution with mean vector $(0,0)$ and covariance matrix $I = \begin{pmatrix} 1 & 0 \\ 0 & 1 \end{pmatrix}$. Show that the squared distance between the points is distributed like $2\chi^2[2]$.

[1.2.16] Show that if $X_1,\cdots,X_n$ are independent and distributed like $N(0,1)$ then $\sum_{i=1}^{n} X_i^2 \sim \chi^2[n]$.

[1.2.17] If X is distributed like $N(\mu,1)$ then the distribution of X^2 is called a **non-central chi-squared**, $\chi^2[1;\lambda]$ with 1 degree of freedom, and parameter of non-centrality $\lambda = \frac{\mu^2}{2}$. Show that the moment generating function of $\chi^2[1;\lambda]$ is equal to

$$M(t) = e^{-\lambda}\sum_{j=0}^{\infty}\frac{\lambda^j}{j!}M_j(t),$$

where $M_j(t) = (1-2t)^{-\frac{1}{2}-j}$, is the m.g.f. of $\chi^2[1+2j]$.

[1.2.18] Show that if Q_1 and Q_2 are independent, $Q_i \sim \chi^2[\nu_i, \lambda_i]$, $i = 1, 2$ then $Q_1 + Q_2 \sim \chi^2[\nu_1 + \nu_2; \lambda_1 + \lambda_2]$.

[1.3.1] Consider Example 1.3, with the parameters $l = 1$, $\sigma = 1$ and $\lambda = 10$. Let $\xi = 1$. What is the probability that the point ξ is covered by at least 2 random intervals?

[1.3.2] Consider Example 1.4. $N = 50$ artillery rounds are aimed at the point $\mathbf{P}_0 = (0, 0)$. The burst points of the rounds follow the bivariate normal distribution with mean $(-10, 20)$ (due to aiming bias) and standard deviations $\sigma_x = 50$ and $\sigma_y = 80$ [meters]. The correlation is $\rho = 0$. If a round bursts within $R = 10$ [meters] from $\mathbf{P}_0$ it destroys it. What is the probability that $\mathbf{P}_0$ will survive?

7.1.2. Solutions of Problems For Chapter 1

[1.1.1] Pr{Point is in $C\bar{S}$} $= 1 - 2/\pi$.

[1.1.2] Without loss of generality, we can assume that the length of the stick is $l = 1$. Let X, Y, Z be the length of the three pieces, where $X + Y + Z = 1$. The sample space is $\Omega = \{(X, Y) : 0 \leq X, Y, X + Y \leq 1\}$. The three pieces of the broken stick can form a triangle if:
(i) $X + Y \geq Z$ or (i) $X + Y \geq \frac{1}{2}$
(ii) $X + Z \geq Y$ or (ii) $Y \leq \frac{1}{2}$
(iii) $Y + Z \geq X$ or (iii) $X \leq \frac{1}{2}$.
The set of (X, Y) points satisfying (i) - (iii) is a triangle whose vertices are at $(\frac{1}{2}, 0)$, $(0, \frac{1}{2})$ and $(\frac{1}{2}, \frac{1}{2})$. The area of this triangle relative to the area of Ω is $1/4$.

[1.1.3] (i) $\Pr\{X + Y \leq 1\} = \frac{7}{8} = 0.875$,
(ii) $\Pr\{X^2 + Y^2 \leq 1\} = \frac{\pi}{4} = 0.785$,
(iii)

$$\begin{aligned}
&\Pr\{-1 \leq X + Y \leq 1\} \mid X^2 + Y^2 \leq 1\} \\
&= \frac{\Pr\{\{-1 \leq X + Y \leq 1\} \cap \{X^2 + Y^2 \leq 1\}}{\Pr\{X^2 + Y^2 \leq 1\}} \\
&= \frac{(1 + \frac{\pi}{2})/4}{\pi/4} = \frac{1}{2} + \frac{1}{\pi} = 0.8183.
\end{aligned}$$

1.1.4] A random disk centered at T does not intersect the line connecting the origin with $(0, 1)$ (the y-axis) if its radius, R, is smaller than $|x|$, where x is the x-coordinate of its center. Thus,

$$\Pr\{R < |X|\} = \int_{-1}^{1} \Pr\{R < |x| \mid X = x\}(1 - |x|)dx,$$

since $f(x) = 1 - |x|$, for $-1 \le x \le 1$, is the p.d.f. of X. Furthermore,

$$\Pr\{R < |x| \mid X = x\} = G(|x| \mid x) = \begin{cases} 1, & \text{if } |x| \ge \frac{1}{2} \\ \dfrac{10|x|^3 - 15|x|^4 + 6|x|^5}{10(1-|x|)^3 - 15(1-|x|)^4 + 6(1-|x|)^5}, & \text{if } |x| < \frac{1}{2} \end{cases}$$

Hence,

$$\Pr\{R < |X|\} = \frac{1}{4} + 2\int_0^{1/2} \frac{10x^3 - 15x^4 + 6x^5}{10(1-x)^2 - 15(1-x)^3 + 6(1-x)^4} dx = 0.3899.$$

[1.2.1] (i) $\Pr\{4 \le X \le 10\} = 0.8694 - 0.0397 = 0.8297$.
(ii) $\Pr\{X > 13\} = 0.0145$.
(iii) $\sum_{a=1}^{45} \Pr\{X \ge a\} = 7.6498$, $E\{X\} = np = 45 \cdot 0.17 = 7.65$.

[1.2.3] Let X_i be the number of obscuring objects in $\mathcal{S}_i$ $(i = 1, 2, 3)$. The number of obscuring objects in $\mathcal{S}$ is $T = X_1 + X_2 + X_3$. T has a Poisson distribution with mean $\lambda_T = 45$

$$\Pr\{T \ge 50\} = 0.2512.$$

[1.2.4] Conditional on $N = n$ (given the number of trees), the probability that the j-th smallest diameter is not larger than x is given by

$$F_{j,n}(x) = \begin{cases} 0, & x \le a \\ 1 - B(j-1; n, F(x)), & a < x < b \\ 1, & x \ge b \end{cases}$$

where $B(k; n, \theta)$ is the c.d.f. of the Binomial distribution. Thus, the c.d.f. of the j-th smallest diameter is the Poisson mixture of $F_{j,n}(x)$, namely

$$F_j(x) = e^{-\lambda} \sum_{n=0}^{\infty} \frac{\lambda^n}{n!} F_{j,n}(x).$$

Hence,

$$F_j(x) = \begin{cases} 0, & x \le a \\ 1 - e^{-\lambda F(x)} \displaystyle\sum_{i=0}^{j-1} \frac{(\lambda F(x))^i}{i!}, & a < x < b \\ 1, & b \le x. \end{cases}$$

[1.2.5] The c.d.f. of the j-th smallest radius, out of n, is $F_{j,n}(x) = 1 - B(j-1; n, F(x))$, where $F(x)$ is the c.d.f. of the radius. In the present problem, $F(x) = \frac{x}{b}$, $0 \le x \le b$. Thus, the conditional expectation of $X_{j,n}$, given $N = n$ is

$$E\{X_{j,n} \mid N = n\} = \int_0^b (1 - F_{j,n}(x))dx$$
$$= \sum_{i=0}^{j} \binom{n}{i} \int_0^b \left(\frac{x}{b}\right)^i \left(1 - \frac{x}{b}\right)^{n-i} dx.$$

Notice that

$$\int_0^b \left(\frac{x}{b}\right)^i \left(1 - \frac{x}{b}\right)^{n-i} dx = b \int_0^1 u^i (1-u)^{n-1} du$$
$$= bB(i+1, n-i+1)$$
$$= b\frac{i!(n-i)!}{(n+1)!}.$$

Hence, for $j \le n$,

$$E\{X_{j,n} \mid N = n\} = b\sum_{i=0}^{j-1} \binom{n}{i} \frac{i!(n-i)!}{n+1}$$
$$= \frac{bj}{n+1}.$$

N has a Poisson distribution with mean λ. Accordingly,

$$E\{X_{j,n} \mid N \ge j\} = bj\frac{\sum_{n=j}^{\infty} \frac{1}{n+1} p(n;\lambda)}{1 - P(j-1;\lambda)}$$
$$= \frac{bj}{\lambda} \cdot \frac{e^{-\lambda} \sum_{n=j}^{\infty} \frac{\lambda^{n+1}}{(n+1)!}}{1 - P(j-1;\lambda)}$$
$$= \frac{bj}{\lambda} \cdot \frac{1 - P(j;\lambda)}{1 - P(j-1;\lambda)}.$$

[1.2.6]

$$I_\theta(k+1, n-k) = \frac{1}{B(k+1, n-k)} \int_0^\theta u^k (1-u)^{u-k-1} du$$
$$B(k+1, n-k) = \frac{1}{(n-k)\binom{n}{k}}.$$

Thus, integration by parts yields

$$I_\theta(k+1, n-k) = -\binom{n}{k}\theta^k(1-\theta)^{n-k} + I_\theta(k, n-k+1).$$

Continuing in this manner we obtain

$$I_\theta(k+1, n-k) = -\sum_{j=1}^{k}\binom{n}{j}\theta^j(1-\theta)^{n-j} + I_\theta(1, n).$$

But

$$I_\theta(1, n) = n\int_0^\theta (1-u)^{n-1}du = 1-(1-\theta)^n.$$

Hence,

$$\begin{aligned} I_\theta(k+1, n-k) &= 1 - \sum_{j=0}^{k}\binom{n}{j}\theta^j(1-\theta)^{n-j} \\ &= 1 - B(k; n, \theta). \end{aligned}$$

[1.2.7]

$$\begin{aligned} \Pr\{G(1, k+1) \le x\} &= \frac{1}{k!}\int_0^x u^k e^{-u}du \\ &= -\frac{1}{k!}u^k e^{-u}\Big|_0^x + \frac{k}{k!}\int_0^x u^{k-1}e^{-u}du \\ &= -\frac{1}{k!}x^k e^{-x} + \frac{1}{(k-1)!}\int_0^x u^{k-1}e^{-u}du \\ &= -\sum_{j=1}^{k}\frac{1}{j!}x^j e^{-x} + \int_0^x e^{-u}du \\ &= 1 - \sum_{j=0}^{k}\frac{1}{j!}x^j e^{-x} = 1 - P(k; x). \end{aligned}$$

[1.2.8] (i) Notice first that X is distributed like $\mu + \sigma N(0, 1)$. Let $M_Z(t)$ denote th m.g.f. of $N(0, 1)$.

$$M_Z(t) = \frac{1}{\sqrt{2\pi}}\int_{-\infty}^{\infty} e^{tx-\frac{1}{2}x^2}dx = e^{t^2/2}\cdot\frac{1}{\sqrt{2\pi}}\int_{-\infty}^{\infty} e^{-\frac{1}{2}(x-t)^2}dx.$$

Thus, $M_Z(t) = e^{t^2/2}$. Furthermore

$$\begin{aligned} M(t) &= E\{e^{tX}\} = E\{e^{t(\mu+\sigma Z)}\} \\ &= e^{t\mu}M_Z(t\sigma) = \exp\left\{t\mu + \frac{1}{2}t^2\sigma^2\right\}. \end{aligned}$$

(ii) $M'(t) = (\mu + t\sigma^2)M(t)$. Hence, $E\{X\} = M'(t)\big|_{t=0} = \mu$.

$$M''(t) = \sigma^2 M(t) + (\mu + t\sigma^2)^2 M(t).$$

Therefore, $E = \{X^2\} = M''(t)\big|_{t=0} = \sigma^2 + \mu^2$. Finally, $V\{X\} = E\{X^2\} - (E\{X\})^2 = \sigma^2$.

[1.2.9]

$$\Pr\{X^2 \le \xi\} = \Pr\{-\sqrt{\xi} \le X \le \sqrt{\xi}\} = 2\Phi(\sqrt{\xi}) - 1.$$

Hence, the p.d.f. of X^2 is

$$\begin{aligned} g(x) &= \frac{d}{dx}\{2\Phi(\sqrt{x}) - 1\} \\ &= \frac{1}{\sqrt{2\pi}} e^{-\frac{1}{2}x} x^{-\frac{1}{2}}, \\ &= \frac{1}{\Gamma(\frac{1}{2})2^{1/2}} x^{-1/2} e^{-x/2}, \quad x \ge 0. \end{aligned}$$

Thus, the distribution of X^2 is like that of $G(2, \frac{1}{2})$ or $\chi^2[1]$.

[1.2.10] The joint p.d.f. of (X, Y) is

$$\begin{aligned} f(x,y) = \frac{1}{2\pi\sigma_x\sigma_y\sqrt{1-\rho^2}} \exp\Bigg\{ &- \frac{1}{2(1-\rho^2)}\Bigg[\left(\frac{x-\xi}{\sigma_x}\right)^2 \\ &- 2\rho \cdot \frac{x-\xi}{\sigma_x} \cdot \frac{y-\eta}{\sigma_y} + \left(\frac{y-\eta}{\sigma_y}\right)^2\Bigg]\Bigg\}, \quad -\infty < x, y < \infty. \end{aligned}$$

The marginal p.d.f. of Y is

$$g(y) = \frac{1}{\sqrt{2\pi}\,\sigma_y}\left\{-\frac{1}{2}\left(\frac{y-\eta}{\sigma_y}\right)^2\right\}, \quad -\infty < y < \infty.$$

The conditional p.d.f. of X, given $\{Y = y\}$ is

$$\begin{aligned} h(x \mid y) &= \frac{f(x,y)}{g(y)} \\ &= \frac{1}{\sqrt{2\pi}\,\sigma_x\sqrt{1-\rho^2}} \exp\Bigg\{ - \frac{1}{2(1-\rho^2)}\Bigg[\left(\frac{x-\xi}{\sigma_x}\right)^2 \\ &\quad 2\rho\frac{x-\xi}{\sigma_x} \cdot \frac{y-\eta}{\sigma_y} + \left(\frac{y-\eta}{\sigma_y}\right)^2\Bigg] + \frac{1}{2}\left(\frac{y-\eta}{\sigma_y}\right)^2\Bigg\} \\ &= \frac{1}{\sqrt{2\pi}\,\sigma_x\sqrt{1-\rho^2}} \exp\left\{ - \frac{1}{2\sigma_x^2(1-\rho^2)}\left[x - \xi - \rho\frac{\sigma_x}{\sigma_y}(y-\eta)\right]^2\right\}. \end{aligned}$$

This proves that the conditional distribution of X, given $\{Y = y\}$, is normal with mean

$$E\{X \mid Y = y\} = \xi + \rho\frac{\sigma_x}{\sigma_y}(y - \eta),$$

and variance

$$V\{X \mid Y = y\} = \sigma_x^2(1-\rho^2).$$

To establish Eq. (1.46) notice that

$$\begin{aligned}\Phi_2(z_1, z_2, \rho) &= \Pr\{Z_1 \le z_1, Z_2 \le z_2\} \\ &= \int_{-\infty}^{z_2} \Pr\{Z_1 \le z_1 \mid Z_2 = x\}\phi(x)dx.\end{aligned}$$

From the above result,

$$\Pr\{Z_1 \le z_1 \mid Z_2 = z_2\} = \Phi\left(\frac{z_1 - \rho z_2}{\sqrt{1-\rho^2}}\right)$$

since $E\{Z_1\} = E\{Z_2\} = 0$ and $V\{Z_1\} = V\{Z_2\} = 1$. Finally, collecting all these results, we obtain

$$\Phi_z(z_1, z_2; \rho) = \frac{1}{\sqrt{2\pi}} \int_{-\infty}^{z_2} e^{-\frac{1}{2}x^2} \Phi\left(\frac{z_1 - \rho x}{\sqrt{1-\rho^2}}\right) dx.$$

[1.2.11]

$$\begin{aligned}|V| &= \sigma_x^2\sigma_y^2 - \sigma_{xy}^2 = \sigma_x^2\sigma_y^2 - \rho^2\sigma_x^2\sigma_y^2 \\ &= \sigma_x^2\sigma_y^2(1-\rho^2). \\ V^{-1} &= \frac{1}{\sigma_x^2\sigma_y^2(1-\rho^2)}\begin{bmatrix}\sigma_y^2 & -\rho\sigma_x\sigma_y \\ -\rho\sigma_x\sigma_y & \sigma_x^2\end{bmatrix} \\ &= \frac{1}{1-\rho^2}\begin{bmatrix}\dfrac{1}{\sigma_x^2} & -\rho\dfrac{1}{\sigma_x\sigma_y} \\ -\dfrac{\rho}{\sigma_x\sigma_y} & \dfrac{1}{\sigma_y^2}\end{bmatrix}.\end{aligned}$$

Thus,

$$(x-\xi, y-\eta)V^{-1}\begin{bmatrix}x-\xi \\ y-\eta\end{bmatrix} = \frac{1}{1-\rho^2}\left[\left(\frac{x-\xi}{\sigma_x}\right)^2 - 2\rho\frac{x-\xi}{\sigma_x}\cdot\frac{y-\eta}{\sigma_y} + \left(\frac{y-\eta}{\sigma_y}\right)^2\right.$$

[1.2.12] We start by making the transformation to $Z_1 = \dfrac{x_1 - \xi}{\sigma_x}$ and $Z_2 = (Y-\eta)/\sigma$. The m.g.f. of (Z_1, Z_2) is

$$\begin{aligned}M^*(t_1, t_2) = \frac{1}{2\pi(1-\rho^2)^{1/2}} \iint \exp\{&t_1 z_1 + t_2 z_2 \\ &- \frac{1}{2(1-\rho^2)}(z_1^2 - 2\rho z_1 z_2 + z_2^2)\}dz_1dz_2.\end{aligned}$$

According to Exercise [1.2.10], the conditional distribution of Z_2, given $Z_1 = z$, is $N(\rho z_1, 1-\rho^2)$. Hence, we can write

$$M^*(t_1,t_2) = \frac{1}{2\pi}\int_{-\infty}^{\infty} e^{t_1 z_1 - \frac{1}{2}z_1^2} \left\{ \frac{1}{\sqrt{2\pi}\sqrt{1-\rho^2}} \int_{-\infty}^{\infty} e^{t_2 z_2 - \frac{1}{2}\frac{1}{1-\rho^2}(z_2-\rho z_1)^2} dz_2 dz_1 \right\}$$

$$\begin{aligned} t_2 z_2 - \frac{1}{2}\cdot\frac{1}{1-\rho^2}[z_2-\rho z_1]^2 &= -\frac{1}{2}\cdot\frac{1}{1-\rho^2}[z_2^2 - 2\rho z_1 z_2 - 2(1-\rho^2)t_2 z_2] \\ &= -\frac{1}{2}\cdot\frac{1}{1-\rho^2}[(z_2-(\rho z_1+(1-\rho^2)t_2))^2] \\ &\quad + \frac{1}{2}(1-\rho^2)t_2^2 + \rho z_1 t_2. \end{aligned}$$

Furthermore,

$$\frac{1}{\sqrt{2\pi}\sqrt{1-\rho^2}}\int_{-\infty}^{\infty} \exp\left\{-\frac{1}{2}\cdot\frac{1}{1-\rho^2}[z_2-(\rho z_1+(1-\rho^2)t_2)]^2\right\} dz_2 = 1.$$

Therefore,

$$\begin{aligned} M^*(t_1,t_2) &= e^{\frac{1}{2}(1-\rho^2)t_2^2}\cdot\frac{1}{2\pi}\int_{-\infty}^{\infty}\exp\left\{t_1 z_1 + \rho z_1 t_2 - \frac{1}{2}z_1^2\right\} dz_1 \\ &= \exp\left\{\frac{1}{2}t_2^2 - \frac{1}{2}\rho^2 t_2^2 + \frac{1}{2}\rho^2 t_2^2 + \rho t_1 t_2 + \frac{1}{2}t_1^2\right\} \\ &= \exp\left\{\frac{1}{2}(t_1^2 + t_2^2 + 2\rho t_1 t_2)\right\}. \end{aligned}$$

Finally,

$$\begin{aligned} M(t_1,t_2) &= E\{\exp\{t_1(\xi+\sigma_x Z_1) + t_2(\eta+\sigma_y Z_2)\}\} \\ &= \exp\{t_1\xi + t_2\eta\}M^*(t_1\sigma_x, t_2\sigma_y) \\ &= \exp\left\{t_1\xi + t_2\eta + \frac{1}{2}(t_1^2\sigma_x^2 + t_2^2\sigma_y^2 + 2\rho t_1 t_2\sigma_x\sigma_y)\right\}. \end{aligned}$$

[1.2.13] The m.g.f. of the bivariate normal distribution, can be written in matrix notation as

$$M_X(t_1,t_2) = \exp\{\boldsymbol{\mu}'\mathbf{t} + \frac{1}{2}\mathbf{t}'V\mathbf{t}\},$$

where $\boldsymbol{\mu}' = (\xi,\eta)$, $\mathbf{t}' = (t_1,t_2)$ and $V = \begin{bmatrix} \sigma_x^2 & \rho\sigma_x\sigma_y \\ \rho\sigma_x\sigma_y & \sigma_y^2 \end{bmatrix}$. Since A is non-singular, $\mathbf{X} = A^{-1}\mathbf{U}$. Thus, the m.g.f. of $\mathbf{U}$, at $\mathbf{t} = (t_1,t_2)'$ is

$$\begin{aligned} M_{\mathbf{u}}(\mathbf{t}) &= E\{\exp\{\mathbf{t}'\mathbf{U}\}\} = E\{\exp\{\mathbf{t}'A\mathbf{x}\}\} \\ &= E\{\exp\{(A'\mathbf{t})'\mathbf{X}\}\} = M_{\mathbf{X}}(A'\mathbf{t}). \end{aligned}$$

Thus,

$$M_{\mathbf{U}}(\mathbf{t}) = \exp\left\{\boldsymbol{\mu}'A'\mathbf{t} + \frac{1}{2}\mathbf{t}'AVA'\mathbf{t}\right\}$$
$$= \exp\left\{(A\boldsymbol{\mu})'\mathbf{t} + \frac{1}{2}\mathbf{t}'(AVA')\mathbf{t}\right\}.$$

This implies that $\mathbf{U}$ is normally distributed, with mean $E\{\mathbf{U}\} = A\boldsymbol{\mu}$ and covariance matrix AVA'.

[1.2.14] (i) If $\rho = 0$ then the joint p.d.f. of X, Y is

$$f(x,y) = \frac{1}{2\pi\sigma_x\sigma_y}\exp\left\{-\frac{1}{2}\left[\left(\frac{x-\xi}{\sigma_x}\right)^2 + \left(\frac{y-\eta}{\sigma_y}\right)^2\right]\right\}$$
$$= \frac{1}{\sqrt{2\pi}\,\sigma_x}\exp\left\{-\frac{1}{2}\left(\frac{x-\xi}{\sigma_x}\right)^2\right\}\cdot\frac{1}{\sqrt{2\pi}}\exp\left\{-\frac{1}{2}\cdot\left(\frac{y-\eta}{\sigma_y}\right)^2\right\}.$$

That is, the joint p.d.f. is the product of the two marginals. Hence, X and Y are independent.

(ii) If X and Y are independent then $f(x,y) = f_X(x)\cdot f_Y(y)$ for **all** (x,y). Or,

$$\frac{1}{\sqrt{1-\rho^2}}\exp\left[-\frac{\rho}{2(1-\rho^2)}\left[\left(\frac{x-y}{\sigma_x}\right)^2\rho + \left(\frac{y-\eta}{\sigma_y}\right)^2\rho - 2\frac{x-\xi}{\sigma_x}\cdot\frac{y-\eta}{\sigma_y}\right]\right\} = 1$$

for **all** (x,y). Obviously, this equality holds for all (x,y) if $\rho = 0$. Recall that $|\rho| < 1$. If $\rho > 0$ then

$$1 = \frac{1}{1-\rho^2}\lim_{\substack{x\to\infty\\ y\to-\infty}}\exp\left\{-\frac{\rho}{2(1-\rho^2)}\left[\frac{(x-y)^2}{\sigma_x^2}\rho + \left(\frac{y-\eta}{\sigma_y}\right)^2\rho - 2\frac{x-\xi}{\sigma_x}\cdot\frac{y-\eta}{\sigma_y}\right.\right.$$

which is a contradiction! Thus, ρ cannot be positive. In a similar fashion we show that ρ cannot be negative. Hence $\rho = 0$.

[1.2.15] Notice that the coordinate (X,Y) of the points are independent standard normal random variables. The squared distance between the points is

$$D^2 = (X_1 - X_2)^2 + (Y_1 - Y_2)^2.$$

$X_1 - X_2$ is distributed like $N(0,2)$. Thus, $(X_1 - X_2)^2$ is distributed like $2\chi^2[1]$. Similarly, $(Y_1 - Y_2)^2$ is distributed like $2\chi^2[1]$. Moreover, the X's and the Y's are independent. Hence $D^2 \sim 2(\chi_1^2[1] + \chi_2^2[1]) \sim 2\chi^2[2]$. In Problem [1.7.9] we have shown that X_1^2 is distributed like $\chi^2[1]$. Since all the n random variables are identically distributed, they are all distributed like $\chi^2[1]$. If X and Y are independent, X distributed like $G(\beta,\nu_1)$ and Y distributed like $G(\beta,\nu_2)$ then $X+Y$ is distributed like $G(\beta,\nu_1+\nu_2)$. This can be proven by realizing that the moment generating function of $X+Y$ is the product of their corresponding moment generating functions. The m.g.f. of X is

$$\begin{aligned} M_x(t) &= \frac{1}{\Gamma(\nu_1)\beta^{\nu_1}} \int_0^\infty e^{tx-x/\beta} x^{\nu_1-1} dx \\ &= \frac{1}{\Gamma(\nu_1)\beta^{\nu_1}} \int_0^\infty x^{\nu_1-1} e^{-x(1-t\beta)/\beta} dx \\ &= (1-t\beta)^{-\nu_1}, \quad \text{for } t < \frac{1}{\beta}. \end{aligned}$$

Hence,

$$\begin{aligned} M_{X+Y}(t) &= (1-t\beta)^{-\nu_1}(1-t\beta)^{-\nu_2} \\ &= (1-t\beta)^{-(\nu_1+\nu_2)}. \end{aligned}$$

Finally, since X_i^2 is distributed like $G(2, \frac{1}{2})$, $\sum_{i=1}^{n} X_i^2$ is distributed like $G(2, \frac{n}{2})$ or like $\chi^2[n]$.

[1.2.17] The m.g.f. of X^2 is

$$\begin{aligned} M(t) = E\{e^{tX^2}\} &= \frac{1}{\sqrt{2\pi}} \int_{-\infty}^{\infty} \exp\left\{tx - \frac{1}{2}(x-\mu)^2\right\} dx \\ &= \frac{1}{\sqrt{2\pi}} \int_{-\infty}^{\infty} \exp\left\{-\frac{(1-2t)}{2}\left(x - \frac{\mu}{1-2t}\right)^2\right\} dx \cdot \exp\left\{\frac{t\mu^2}{1-2t}\right\} \\ &= e^{t\mu^2/(1-2t)}(1-2t)^{-1/2}. \end{aligned}$$

Moreover,

$$\frac{2t}{1-2t} = -1 + \frac{1}{1-2t}.$$

Hence

$$e^{t\mu^2/(1-2t)} = e^{-\mu^2/2 + \frac{\mu^2}{2(1-2t)}} = e^{-\mu^2/2} \cdot e^{\frac{\mu^2}{2(1-2t)}}.$$

Taylor series expansion yields

$$e^{\frac{\mu^2}{2(1-2t)}} = \sum_{j=0}^{\infty} \frac{(\mu^2/2)^j}{j!}(1-2t)^{-j}.$$

Substituting above, we obtain

$$\begin{aligned} M(t) &= e^{-\mu^2/2} \sum_{j=0}^{\infty} \frac{\mu^2/2)^j}{j!}(1-2t)^{-j-\frac{1}{2}} \\ &= e^{-\lambda} \sum_{j=0}^{\infty} \frac{\lambda^j}{j!} M_j(t). \end{aligned}$$

Notice that

$$M(t) = E\{(1-2t)^{-\frac{1}{2}-J}\}$$

where J has a Poisson distribution with mean λ.

[1.2.18] Let us start with $\nu_1 = \nu_2 = 1$. Since Q_1 and Q_2 are independent, the m.g.f. of the sum $W = Q_1 + Q_2$ is

$$\begin{aligned} M_W(t) &= M_{Q_1}(t)M_{Q_2}(t) \\ &= E\{(1-2t)^{-\frac{1}{2}-J_1}\}E\{(t-2t)^{-\frac{1}{2}-J_2}\} \end{aligned}$$

where J_1 and J_2 are independent random variables having Poisson distribution with mean λ_1 and λ_2. This independence implies that

$$M_W(t) = E\{(1-2t)^{-1-(J_1+J_2)}\}.$$

But J_1+J_2 is distributed like Poisson with mean $\lambda_1+\lambda_2$. Hence W is distributed like $\chi^2[2;\lambda_1+\lambda_2]$. By induction, we show that, for any integers ν, the m.g.f. of $\chi^2[\nu;\lambda]$ is $M(t) = E\{(1-2t)^{-\frac{\nu}{2}-J}\}$, where J has a Poisson distribution with mean λ. Finally, the m.g.f. of $Q_1 + Q_2$ is

$$\begin{aligned} M(t) &= E\{(1-2t)^{-\frac{\nu_1}{2}-J_1}\}E\{(1-2t)^{-\frac{\nu_2}{2}-J_2}\} \\ &= E\{(1-2t)^{-\frac{1}{2}(\nu_1+\nu_2)-(J_1+J_2)}\}. \end{aligned}$$

Thus, $Q_1 + Q_2$ is distributed like $\chi^2[\nu_1+\nu_2;\lambda_1+\lambda_2]$.

[1.3.1] Equation (1.49), for $l = 1$, $\sigma = 1$, $\xi = 1$ and $\lambda = 10$ yields $\mu(1) = 3.033$. The probability that the point ξ is covered by at least 2 random intervals is $1-\exp\{-\mu(1)\}-\mu(1)\exp\{-\mu(1)\} = 0.7575$.

[1.3.2] The probability that an artillery round destroys $\mathbf{P}_0$ is given by Eq. (1.50), in which $R = 10$, $\sigma_1 = 50$, $\sigma_2 = 80$, $\rho = 0$. Due to the aiming bias, $\xi_1 = 10$ and $\xi_2 = -20$. Thus, Eq. (1.50) yields

$$\begin{aligned} d(\xi_1,\xi_2) &= \frac{1}{50}\int_0^{20}\phi\left(\frac{x}{50}\right)\left[\Phi\left(\frac{-20+(100-(x-10)^2)^{1/2}}{80}\right)\right. \\ &\quad \left. -\ \Phi\left(\frac{-20-(100-(x-10)^2)^{1/2}}{80}\right)\right]dx \\ &= 0.011797. \end{aligned}$$

Finally, the probability that the point $\mathbf{P}_0$ will survive is $(1-0.011797)^{50} = 0.5525$.

7.2.1. Problems For Chapter 2

[2.1.1] Compute $\Pr\{L \text{ intersects } \mathcal{L}\}$ when $\mathcal{L}$ is the y-axis. X_0 has a Cauchy distribution, with p.d.f.. $f(x) = \dfrac{1}{\pi(1+x^2)}$, $-\infty < x < \infty$, and θ has a uniform distribution on $(-\frac{\pi}{2},\frac{\pi}{2})$. X_0 and θ are independent.

[2.2.1] Consider a circle $\mathcal{C}$ centered at the origin, with radius $R = 1$. Let the intercept A, and the slope, B, of a random line be independent standard normal random variables. Show that

$$\Pr\{\mathcal{L} \text{ intersects } \mathcal{C}\} = 0.795.$$

[2.2.2] Show that if $R = 1$, A and B are independent standard normal random variables, then the expected length of the intersecting cord is $E\{D\} = 1.3396$.

[2.2.3] In continuation of the previous problem, compute the variance of D.

[2.3.1] Consider the line $\mathcal{L} : y = 1 + x$. A disk of radius R is placed at random so that, its center (X, Y) has a bivariate standard normal distribution, with zero correlation. The radius, R, is independent of (X, Y) and has a uniform distribution on $(0, 1)$.
(i) Find the probability that the random disk intersects $\mathcal{L}$.
(ii) How many such disks should be independently placed so that the probability of at least one intersection will be at least .95?

[2.4.1] Let $\mathcal{C}$ be a circle of length (of circumference) $L = 1$. Let A be an arc of fixed length $a = \dfrac{1}{2\pi}$. One needs at least 7 such arcs to cover $\mathcal{C}$. Suppose that $n = 10$ such arcs are randomly placed on $\mathcal{C}$. What is the coverage probability? What would be this probability if $n = 20$?

[2.4.2] Consider a circle $\mathcal{C}$, centered at the origin, of radius $R = 1$. Let (ρ, θ) be the polar coordinates of a point $\mathbf{P}$, with $\rho > 1$. From $\mathbf{P}$ draw two tangential lines to $\mathcal{C}$. Let A be the length of the arc on $\mathcal{C}$, between the two points at which these two tangential lines touch $\mathcal{C}$.
(i) Show that $A = \pi - 2\sin^{-1}(\frac{1}{\rho})$.
(ii) Derive the c.d.f. of A, when ρ has an exponential distribution

$$F(\rho) = 1 - \exp\{-(\rho - 1)\}, \quad \rho \geq 1.$$

[2.4.3] In continuation of the previous problem, let $\mathbf{P}_1$ and $\mathbf{P}_2$ be two points, independently chosen (at random) outside $\mathcal{C}$, i.e., $\rho_1 > 1$ and $\rho_2 > 1$. Show that the corresponding arcs on $\mathcal{C}$, which these points generate, are disjoint if,

$$|\theta_2 - \theta_1| > \pi - \left(\sin^{-1}\left(\frac{1}{\rho_2}\right) + \sin^{-1}\left(\frac{1}{\rho_1}\right)\right).$$

[2.5.1] Derive Eq. (2.19).

[2.5.2] Consider $n = 3$ points on a circle $\mathcal{C}$, centered at the origin with radius $R = 1$. The orientation angles of the specified points are, 0, $\frac{\pi}{6}$, $\frac{5\pi}{6}$. An arc is placed at random on $\mathcal{C}$. The arc coordinates (X, Y) are independent random variables. $X \sim \mathcal{U}(0, 2\pi)$ and $Y \sim \text{Beta}(2, 2)$. What is the probability that the three points are simultaneously vacant?

[2.6.1] Show that a disk of radius R centered at (ρ, θ) covers the point $\mathbf{P}_0 = (\rho_0, \theta_0)$ if $(\rho^2 + \rho_0^2 - 2\rho\rho_0 \cos(\theta - \theta_0))^{1/2} < R$.

[2.6.2] What is the probability that the origin is covered by a random disk D, whose center (ρ, θ) is independent of its radius R, $R \sim \mathcal{U}(0,4)$ and the p.d.f. of (ρ, θ) is

$$h(\rho, \theta) = \frac{1}{2\pi} e^{-\rho}, \quad 0 < \rho < \infty, \quad 0 \le \theta \le 2\pi.$$

[2.6.3.] Consider the point $\mathbf{P}_0 = (1,1)$. Let D be a random disk with center coordinate (X, Y) having a bivariate normal distribution centered at the origin, with equal standard deviations $\sigma_x = \sigma_y = 2$. Suppose that the disk radius is fixed at $r = 1$. Apply Eq. (2.27) to find the probability that $\mathbf{P}_0$ is vacant.

[2.6.4] A random disk of radius R is centered at (X, Y). (X, Y) has a bivariate normal distribution with means $(0,0)$ and covariance matrix $\Sigma = \begin{pmatrix} 1 & 0.5 \\ 0.5 & 1 \end{pmatrix}$. The disk radius R is independent of (X, Y) and has a uniform distribution on $(0,2)$. Consider the fixed point $\mathbf{P}_0 = (1,1)$. Show that the probability that $\mathbf{P}_0$ is covered by D is

$$\begin{aligned} &\Pr\{\mathbf{P}_0 \text{ is covered by } D\} \\ &= \frac{1}{2}\int_0^2 \int_{1-R}^{1+R} \phi(x)\left[\Phi\left(\frac{1+(R^2-(x-1)^2)^{1/2}-0.5x}{\sqrt{3}/2}\right)\right. \\ &\left. - \Phi\left(\frac{1-(R^2-(x-1)^2)^{1/2}-0.5x}{\sqrt{3}/2}\right)\right] dx dR \\ &\cong 0.26595. \end{aligned}$$

[2.6.5] Let (X, Y) have a bivariate normal distribution with zero means, and covariance matrix $\Sigma = \begin{pmatrix} 1 & 0.5 \\ 0.5 & 1 \end{pmatrix}$. Determine the probability that (X, Y) belongs to the triangle T, whose vertices are $(1, -2)$, $(2, 3)$ and $(-1, 1)$.

[2.6.6] In continuation of Problem [2.6.5], compute the probability that the triangle T is completely vacant, when the disk radius R is independent of its center (X, Y), and has a uniform distribution on $(0,2)$.

7.2.2. Solution of Problems For Chapter 2

[2.1.1] The c.d.f. of the specified Cauchy distribution is

$$F(x) = \frac{1}{2} + \frac{1}{\pi} \tan^{-1}(x), \quad -\infty < x < \infty.$$

According to Eq. (2.1),

$$\Pr\{L \text{ intersects } \mathcal{L}\} = \frac{1}{\pi} \int_{-\pi/2}^{\pi/2} \left[F\left(\frac{l}{2}\cos(\theta)\right) - F\left(-\frac{l}{2}\cos(\theta)\right)\right] d\theta.$$

Since $F(x)$ is symmetric around $x = 0$,

$$F\left(\frac{l}{2}\cos(\theta)\right) - F\left(-\frac{l}{2}\cos(\theta)\right) = \frac{2}{\pi}\tan^{-1}\left(\frac{l}{2}\cos(\theta)\right).$$

Hence

$$\begin{aligned}\Pr\{L \text{ intersects } \mathcal{L}\} &= \frac{2}{\pi^2}\int_{-\pi/2}^{\pi/2} \tan^{-1}\left(\frac{l}{2}\cos(\theta)\right) d\theta \\ &= \frac{4}{\pi^2}\int_{0}^{\pi/2} \tan^{-1}\left(\frac{l}{2}\cos(\theta)\right) d\theta \\ &= \frac{8}{l\pi^2}\int_{0}^{l/2} \frac{\tan^{-1}(y)}{\sqrt{1-\frac{4}{l^2}y^2}}dy.\end{aligned}$$

This integral can be evaluated numerically. For $l = 1$ we obtain $\Pr\{L$ intersects $\mathcal{L}\} = 0.1926$. For $l = 5$ we obtain $\Pr\{\cdot\} = 0.583$.

[2.2.1] Since A and B are independent, $F_{A|B}(R\sqrt{1+b^2} \mid b) = \Phi(R\sqrt{1+b^2})$. Thus, Eq. (2.2) assumes the form, when $R = 1$,

$$\Pr\{\mathcal{L} \text{ intersects } \mathcal{C}\} = 2\int \phi(z)\Phi(\sqrt{1+z^2})dz - 1.$$

Numerical evaluation of this integral yields the value of 0.795.

[2.2.2] The expected value of D, when $R = 1$, is

$$E\{D\} = \int_0^2 [1 - \Pr\{D \le x\}]dx.$$

Eq. (2.2.6) and the assumption that A, B are independent standard normal yield

$$\begin{aligned}E\{D\} &= 2\left\{\int_0^2\int_{-\infty}^{\infty}\phi(z)\Phi\left(\left(1-\left(\frac{x}{2}\right)^2\right)^{1/2}(1+z^2)^{1/2}\right)dzdx - 1\right\} \\ &= 4\int_0^1 \frac{y}{(1-y^2)^{1/2}}\int_{-\infty}^{\infty}\phi(z)\Phi(y(1+z^2)^{1/2})dzdy - 2 = 1.3396.\end{aligned}$$

[2.2.3] The second moment of D is

$$\begin{aligned}E\{D^2\} &= 2\int_0^2 x(1-\Pr\{D \le x\})dx \\ &= 2\int_0^2 x\left[2\int_{-\infty}^{\infty}\phi(z)\Phi\left(\left(1-\left(\frac{x}{2}\right)^2\right)^{1/2}(1+z^2)^{1/2}\right)dz - 1\right]dx \\ &= 16\int_0^1 y\int_{-\infty}^{\infty}\phi(z)\Phi(y(1+z^2)^{1/2})dzdy - 4 \\ &= 2.3734.\end{aligned}$$

Accordingly, the variance of D is

$$V\{D\} = 2.3734 - (1.3396)^2 = 0.5789.$$

[2.3.1] A disk of radius $R = r$ intersects $\mathcal{L}$ if, and only if, its center is between the lines $\mathcal{L}_r^+$ and $\mathcal{L}_r^-$, parallel to $\mathcal{L}$, of distance r from $\mathcal{L}$; on its two sides. The equation of these lines are:

$$\mathcal{L}_r^- : y = 1 - r\sqrt{2} + x$$

and

$$\mathcal{L}_r^+ : y = 1 + r\sqrt{2} + x.$$

Hence,

$$\Pr\{D \text{ intersects } \mathcal{L}\} = \Pr\{1 - \sqrt{2}\,R < Y - X < 1 + \sqrt{2}\,R\}.$$

$Y - X$ has a normal distribution with mean 0 and standard deviation $\sqrt{2}$. Thus, since X, Y and R are independent,

$$\Pr\{D \text{ intersects } \mathcal{L}\} = E\left\{\Phi\left(\frac{1}{\sqrt{2}} + R\right) - \Phi\left(\frac{1}{\sqrt{2}} - R\right)\right\}.$$

Numerical integration yields the value 0.298018. Notice that the above expectation can be approximated in the following manner. Taylor's expansion around $\frac{1}{\sqrt{2}}$ yields

$$\Phi\left(\frac{1}{\sqrt{2}} + R\right) \cong \Phi\left(\frac{1}{\sqrt{2}}\right) + R\phi\left(\frac{1}{\sqrt{2}}\right) - \frac{R^2}{2\sqrt{2}}\phi\left(\frac{1}{\sqrt{2}}\right) - \frac{R^3}{12}\phi\left(\frac{1}{\sqrt{2}}\right)$$

$$\Phi\left(\frac{1}{\sqrt{2}} - R\right) \cong \Phi\left(\frac{1}{\sqrt{2}}\right) - R\phi\left(\frac{1}{\sqrt{2}}\right) - \frac{R^2}{2\sqrt{2}}\phi\left(\frac{1}{\sqrt{2}}\right) + \frac{R^3}{12}\phi\left(\frac{1}{\sqrt{2}}\right)$$

where $\phi(z)$ is the standard normal p.d.f. Thus,

$$\Phi\left(\frac{1}{\sqrt{2}} + R\right) - \Phi\left(\frac{1}{\sqrt{2}} - R\right) \cong 2R\phi\left(\frac{1}{\sqrt{2}}\right) - \frac{R^3}{6}\phi\left(\frac{1}{\sqrt{2}}\right).$$

and

$$\begin{aligned}\Pr\{D \text{ intersects } \mathcal{L}\} &\cong \phi\left(\frac{1}{\sqrt{2}}\right)\left\{2E\{R\} - \frac{1}{6}E\{R^3\}\right\} \\ &= \phi\left(\frac{1}{\sqrt{2}}\right)\frac{23}{24} = 0.29775.\end{aligned}$$

(ii) The probability that a disk does not intersect $\mathcal{L}$ is 0.702. We should find the smallest n such that $1 - 0.702^n \geq 0.95$. Hence, $n \geq \dfrac{\log 0.05}{\log 0.702} = 8.47$. We need 9 disks.

[2.4.1] According to Stevens' formula (Eq. (2.18))

$$\Pr\{\text{Coverage}\} = \sum_{l=0}^{6}(-1)^l\binom{10}{l}\left(1 - \frac{l}{2\pi}\right)^9 = 0.00397$$

since $k = [\frac{1}{a}] = [2\pi] = 6$. If the number of arcs is $n = 20$ the coverage probability is

$$\Pr\{\text{Coverage}\} = \sum_{l=0}^{6}(-1)^l \binom{20}{l}\left(1 - \frac{l}{2\pi}\right)^{19} = 0.3835.$$

[2.4.2] (i) The angle α between the line connecting $\mathbf{P}$ with the origin $\mathbf{O}$ and the tangential line from $\mathbf{P}$ to $\mathcal{C}$ is $\alpha = \sin^{-1}(\frac{1}{\rho})$. The length of the arc between the two tangential lines is therefore $A = 2(\frac{\pi}{2} - \alpha) = \pi - 2\sin^{-1}(\frac{1}{\rho})$.

(ii) Let $G(\lambda)$ be the c.d.f. of A.

$$\begin{aligned} G(\lambda) &= \Pr\{A \le \lambda\} \\ &= \Pr\left\{\pi - 2\sin^{-1}\left(\frac{1}{\rho}\right) \le \lambda\right\} \\ &= P\left\{\sin^{-1}\left(\frac{1}{\rho}\right) \ge \frac{\pi - \lambda}{2}\right\} \\ &= P\left\{\frac{1}{\rho} \ge \sin\left(\frac{\pi-\lambda}{2}\right)\right\} \\ &= P\left\{\rho \le \frac{1}{\sin\left(\frac{\pi-\lambda}{2}\right)}\right\} \\ &= 1 - e\cdot\exp\left\{-\frac{1}{\sin\left(\frac{\pi-\lambda}{2}\right)}\right\}, \quad 0 \le \lambda \le \pi. \end{aligned}$$

Notice that

$$\begin{aligned} E\{A\} &= e\int_0^{\pi}\exp\left\{-\frac{1}{\sin\left(\frac{\pi-\lambda}{2}\right)}\right\}d\lambda \\ &= 1.785. \end{aligned}$$

[2.4.3] Without loss of generality, assume that $\theta_1 = 0$. The endpoints of the arc generated by $\mathbf{P}_1$, have orientation angles

$$w_{1,2} = \pm\left(\frac{\pi}{2} - \sin^{-1}\left(\frac{1}{\rho_1}\right)\right).$$

Suppose that $\theta_2 > 0$. The arc generated by $\mathbf{P}_2$ has endpoints with orientation angles $\theta_2 \pm \left(\frac{\pi}{2} - \sin^{-1}\left(\frac{1}{\rho_2}\right)\right)$. The two arcs are disjoint if $\theta_2 -$

$$\left(\frac{\pi}{2}-\sin^{-1}\left(\frac{1}{\rho^2}\right)\right) > \frac{\pi}{2}-\sin^{-1}\left(\frac{1}{\rho_1}\right), \text{ or } \theta_2 > \pi-\left(\sin^{-1}\left(\frac{1}{\rho_1}\right)+\sin^{-1}\left(\frac{1}{\rho_2}\right)\right)$$

If $\theta_2 < 0$, then, the two arcs are disjoint if

$$\theta_2 + \left(\frac{\pi}{2}-\sin^{-1}\left(\frac{1}{\rho_2}\right)\right) < -\left(\frac{\pi}{2}-\sin^{-1}\left(\frac{1}{\rho_1}\right)\right)$$

or

$$\theta_2 < -\pi + \left(\sin^{-1}\left(\frac{1}{\rho_1}\right)+\sin^{-1}\left(\frac{1}{\rho_2}\right)\right).$$

Thus, the condition for disjoint arcs is

$$|\theta_2| > \pi - \left(\sin^{-1}\left(\frac{1}{\rho_1}\right)+\sin^{-1}\left(\frac{1}{\rho_2}\right)\right).$$

Generally, if $\theta_1 \neq 0$, we have

$$|\theta_2-\theta_1| > \pi - \left(\sin^{-1}\left(\frac{1}{\rho_1}\right)+\sin^{-1}\left(\frac{1}{\rho_2}\right)\right).$$

[2.5.1] Let $y > t$. The arc from $(1,0)$ to $(1,t)$ is completely vacant if $x > t$ and $X+Y < 2\pi - t$. Hence,

$$\begin{aligned} q(t) &= \frac{1}{2\pi}\int_t^{2\pi} F_Y(2\pi - x)dx \\ &= \frac{1}{2\pi}\int_0^{2\pi-t} F_Y(x)dx. \end{aligned}$$

Notice that $\psi(0) = E\{Y\}$. Hence

$$\psi(t) = E\{Y\} - t + \int_0^t F_Y(x)dx,$$

or

$$\psi(2\pi - t) = E\{Y\} - (2\pi - t) + \int_0^{2\pi-t} F_Y(x)dx.$$

Hence

$$\begin{aligned} q(t) &= \frac{1}{2\pi}[2\pi - t - E\{Y\} + \psi(2\pi - t)] \\ &= 1 - \frac{1}{2\pi}(t + E\{Y\} - \psi(2\pi - t)). \end{aligned}$$

[2.5.2] $t_0 = \frac{7}{6}\pi$, $t_1 = \frac{1}{6}\pi$, $t_2 = \frac{4}{6}\pi$, $E\{Y\} = \frac{1}{2}$. The c.d.f. of Y is

$$F_Y(y) = \begin{cases} 0, & y \le 0 \\ y^2(3-2y), & 0 < y \le 1 \\ 1, & 1 \le y. \end{cases}$$

Accordingly,

$$\psi(t) = \int_t^{2\pi} [1 - F_Y(y)]dy$$
$$= \begin{cases} \frac{1}{2}(1-t)(1-t-t^2+t^3), & t < 1 \\ 0, & 1 \le t. \end{cases}$$

Thus, $\psi(t_0) = 0$, $\psi(t_1) = 0.33758$ and $\psi(t_2) = 0.03086$. Finally, according to Eq. (2.21)

$$p(t_1, t_2) = 1 - \frac{3}{2\pi} \cdot \frac{1}{2} + \frac{1}{2\pi}(\psi(t_0) + \psi(t_1) + \psi(t_2))$$
$$= 0.8199.$$

[2.6.1] Let $\mathbf{v}$ be the vector connecting $\mathbf{O}$ with the disk center and $\mathbf{v}_1$ the vector connecting $\mathbf{O}$ with $\mathbf{P}_0$. $|\mathbf{v}| = \rho$ and $|\mathbf{v}_1| = \rho_0$. The angle between $\mathbf{v}$ and $\mathbf{v}_0$ is $(\theta - \theta_0)$. It follows that the squared distance between the disk center and $\mathbf{P}_0$ is $|\mathbf{v}|^2 + |\mathbf{v}_0|^2 - 2|\mathbf{v}||\mathbf{v}_0|\cos(\theta - \theta_0)$. Hence, the disk covers $\mathbf{P}_0$ only if

$$R^2 > \rho^2 + \rho_0^2 - 2\rho\rho_0\cos(\theta - \theta_0),$$

or

$$R > (\rho^2 + \rho_0^2 - 2\rho\rho_0\cos(\theta - \theta_0))^{1/2}.$$

[2.6.2] A disk of radius R covers the origin if $\{\rho < R\}$. Hence

$$\Pr\{\mathbf{O} \text{ is covered by } D\} = \Pr\{\rho < R\}$$
$$= 1 - E\{e^{-R}\}$$
$$= 1 - \frac{1}{4}\int_0^4 e^{-x}dx = 1 - \frac{1}{4}(1 - e^{-4})$$
$$= 0.7546.$$

[2.6.3] When $\sigma_x = \sigma_y = 2$, the distribution of $(X - x_0)^2 + (Y - y_0)^2$ is like that of $4\chi^2[2;\mu]$, where $\mu = \frac{1}{2\sigma^2}(x_0^2 + y_0^2) = \frac{1}{4}$. Thus,

$$\Pr\{(X-1)^2 + (Y-1)^2 > 1\} = \Pr\{4\chi^2[2;\mu] > 1\}$$
$$= \Pr\left\{\chi^2\left[2;\frac{1}{4}\right] > \left(\frac{1}{2}\right)^2\right\}$$
$$= e^{-1/4}\sum_{j=0}^{d} \frac{\left(\frac{1}{4}\right)^j}{j!}\Pr\left\{X^2[2+2j] \ge \frac{1}{4}\right\}$$
$$= e^{-1/4}\sum_{j=0}^{\infty} \frac{\left(\frac{1}{4}\right)^j}{j!}P\left(j;\frac{1}{8}\right)$$
$$= 0.90708.$$

[2.6.4] The random disk covers $\mathbf{P}_0$ if $(X-1)^2+(Y-1)^2 \le R^2$. This event occurs only if $1-R < X < 1+R$, and $1-(R^2-(X-1)^2)^{1/2} < Y < 1+(R^2-(X-1)^2)^{1/2}$. The conditional distribution Y given X is normal with mean ρX and variance $(1-\rho^2)$. Hence

$$\begin{aligned}\Pr\{\mathbf{P}_0 \text{ is covered}\} &= \Pr\{1-R < X < 1+R, 1-(R^2-(X-1)^2)^{1/2} \\ &\quad < Y < 1+(R^2-(X-1)^2)^{1/2}\} \\ &= \frac{1}{2}\int_0^2 \int_{1-R}^{1+R} \phi(x)\left[\Phi\left(\frac{1+(R^2-(X-1)^2)^{1/2}-0.5x}{\sqrt{3}/2}\right)\right. \\ &\quad \left. -\Phi\left(\frac{1-(R^2-(X-1)^2)^{1/2}-0.5x}{\sqrt{3}/2}\right)\right] dx dR = 0.26595.\end{aligned}$$

[2.6.5] The vertices of the triangle T are $\mathbf{P}_0 = (1,-2)$, $\mathbf{P}_1 = (2,3)$, $\mathbf{P}_2 = (-1,1)$. Thus, we make the orthogonal transformation, with the matrix

$$A = \frac{1}{(1+B^2)^{1/2}}\begin{bmatrix} 1 & B \\ -B & 1\end{bmatrix},$$

with $B = \frac{2}{3}$, or

$$A = \frac{3}{\sqrt{13}}\begin{pmatrix} 1 & \frac{2}{3} \\ -\frac{2}{3} & 1\end{pmatrix}.$$

The transformed vertices are

$$\mathbf{P}'_0 = \left(-\frac{1}{\sqrt{13}}, -\frac{8}{\sqrt{13}}\right), \quad \mathbf{P}'_1 = \left(\frac{12}{\sqrt{13}}, \frac{5}{\sqrt{13}}\right), \quad P'_2 = \left(-\frac{1}{\sqrt{13}}, \frac{5}{\sqrt{13}}\right).$$

The transformed disk centers (X', Y') have a bivariate normal distribution with means $(0,0)$ and covariance matrix

$$\begin{aligned}\Sigma' = A\begin{pmatrix} 1 & 0.5 \\ 0.5 & 1\end{pmatrix} A' &= \frac{1}{26}\begin{pmatrix} 38 & 5 \\ 5 & 14\end{pmatrix} \\ &= \begin{pmatrix} 1.46154 & 0.19231 \\ 0.19231 & 0.53846\end{pmatrix}.\end{aligned}$$

Thus, the probability that (X', Y') is within the transformed triangle is

$$\Pr\left\{-\frac{1}{\sqrt{13}} < X' < \frac{12}{\sqrt{13}}, X' - \frac{7}{\sqrt{13}} < Y' < \frac{5}{\sqrt{13}}\right\}.$$

The conditional distribution of Y', given X', is normal with mean $0.13158x$

and variance 0.5131558. Thus,

$$\Pr\{(X',Y') \in T'\} = \frac{1}{\sqrt{2\pi}\cdot 1.20894}\int_{-\frac{1}{\sqrt{13}}}^{\frac{12}{\sqrt{13}}} \exp\left\{-\frac{1}{2.92308}x^2\right\} \cdot \left[\Phi\left(\frac{\frac{5}{\sqrt{13}} - 0.13158x}{0.716349}\right) - \Phi\left(\frac{x - \frac{7}{\sqrt{13}} - 0.13158x}{0.716349}\right)\right] dx = 0.78209.$$

[2.6.6] We have seen in the previous problem that the transformed triangle has vertices $\mathbf{P}_0' = (-\frac{1}{\sqrt{13}}, -\frac{8}{\sqrt{13}})$, $\mathbf{P}_1' = (\frac{12}{\sqrt{13}}, \frac{5}{\sqrt{13}})$ and $\mathbf{P}_2' = (-\frac{1}{\sqrt{13}}, \frac{5}{\sqrt{13}})$. Using the results of the previous exercise, we obtain: For a given disk radius R, the extended triangle Δ_R has vertices $\mathbf{P}_0^+ = (-\frac{1}{\sqrt{13}} - R, -\frac{8}{\sqrt{13}} - (1+\sqrt{2})R)$, $\mathbf{P}_1^+ = (\frac{12}{\sqrt{13}} + \sqrt{2}R, \frac{5}{\sqrt{13}} + R)$ and $\mathbf{P}_2^+ = (-\frac{1}{\sqrt{13}} - R, \frac{5}{\sqrt{13}} + R)$. The conditional probability that the disk center (X',Y') is with Δ_R is

$$\Pr\{(X',Y') \in \Delta_R \mid R\} = \frac{1}{1.28094}\int_{-\frac{1}{\sqrt{13}}-R}^{\frac{12}{\sqrt{13}}+\sqrt{2}R} \phi\left(\frac{x}{1.20894}\right) \left[\Phi\left(\frac{\frac{5}{\sqrt{13}} + R - 0.13158x}{0.716349}\right) - \Phi\left(\frac{-\frac{7}{\sqrt{13}} - \sqrt{2}R + x - 0.13158x}{0.716349}\right)\right] dx.$$

Finally, the probability that T is completely vacant is

$$1 - \frac{1}{2}\int_0^2 \Pr\{(X',Y') \in \Delta_R \mid R\} dR = 0.1988.$$

7.3.1. Problems for Chapter 3

[3.1.1] Consider a standard Poisson field of random disks, with intensity $\lambda = .01[1/\text{m}^2]$. Let Y designate the radius of a random disk, and suppose that the distribution of Y/b is Beta(2, 2), where $b = 1.5$[m].
(i) What is the distribution of the number of random disks intersecting the line segment between **O** and **T**, when both **O** and **T** are in the field, the distance between **O** and **T** is $d = 100$[m], and **O** is not covered?
(ii) What is the probability that the point **T** is visible from **O**?

[3.1.2] What is the conditional distribution of the length L of a line of sight from **O**, along a ray $\mathbb{R}_s$, with orientation s, in a standard Poisson field of intensity $\lambda[1/\text{m}^2]$, when the radii of disks are uniformly distributed on (1, 2)[m], given that the origin **O** is not covered? Compute the conditional probability that the length L of the line of sight will be at least 30[m], when $\lambda = .1[1/\text{m}^2]$.

[3.1.3] Consider a linear target which is 5[m] long. The target is within a standard Poisson field with $\lambda = 0.02[1/\text{m}^2]$ and disk radius having a uniform distribution on (0.10,0.30)[m]. How far could one be from the target so that, a segment of at least 1[m] of the target will be completely visible, with probability greater than 0.8? [Apply formula (3.5).]

[3.1.4] Let $\mathbf{T}_i$ $(i = 1, \cdots, 4)$ be four target points within a forest. It is assumed that trees are scattered according to a standard Poisson field, with an intensity of $\lambda = .01[1/\text{m}^2]$. The distances of the targets from the observation point $\mathbf{O}$ are: $r_1 = 75[\text{m}]$, $r_2 = 50[\text{m}]$, $r_3 = 35[\text{m}]$ and $r_4 = 100[\text{m}]$. Their orientation angles are 15°, 5°, −5°, −25°, respectively. The radii of tree trunks 1[m] above the ground are identically distributed, with p.d.f.

$$f(y) = \begin{cases} 0.648 \exp\{-(y - .75)/2.5\}, & \text{if } .75 < y < 1.5 \\ 0, & \text{otherwise.} \end{cases}$$

Compute:
(i) The simultaneous visibility probability of $\mathbf{T}_1$ and $\mathbf{T}_4$.
(ii) The simultaneous visibility probability of $\mathbf{T}_1$, $\mathbf{T}_3$, $\mathbf{T}_4$.
(iii) The simultaneous visibility probability of all the four targets.

[3.1.5] Recompute Example 3.5, with parameters, $u = 25[\text{m}]$, $\lambda = 0.1[1/\text{m}^2]$, and the distribution of Y has mean of 0.75[m].

[3.1.6] Let $\mathcal{S}$ be a scattering strip with horizontal boundaries at distances $u = 25[\text{m}]$ and $w = 100[\text{m}]$. The scattering is according to a standard Poisson field with intensity $\lambda = .001[1/\text{m}^2]$. Three target points are on the other side of $\mathcal{S}$, with respect to $\mathbf{O}$. Their orientation angles are $s_1 = 75°$, $s_2 = 15°$ and $s_3 = -5°$. The radii of trees are uniformly distributed over [.5,1.0]. Compute the simultaneous visibility probabilities of the three targets, when $u = 25, 50, 75[\text{m}]$. [Use program SIMVP.]

[3.1.7] Let $\mathcal{S}$ be a scattering strip with horizontal boundaries at $u = 50[\text{m}]$ and $w = 100[\text{m}]$, with an additional rectangle, having coordinates (-20,100), (20,100), (20,150), (-20,150). The scattering within $\mathcal{S}$ is according to a standard Poisson field with intensity $\lambda = .005[1/\text{m}^2]$. The radii of trees are uniformly distributed on (1.,2.)[m]. Compute the simultaneous visibility probability of $k = 4$ targets having orientation angles 45°, 5°, −5°, −45°.

[3.1.8] Let $\mathcal{S}$ be a scattering region consisting of two rectangular regions $\mathcal{S}_1$ and $\mathcal{S}_2$. The boundaries of $\mathcal{S}_1$ are (in rectangular coordinates)

$$u_1(x) = 50 + .25x, \qquad -200 \le x \le 0$$

and

$$w_1(x) = 100 + .25x, \qquad -400 \le x \le 100.$$

The boundaries of $\mathcal{S}_2$ are

$$u_2(x) = 50 - .5x, \qquad 0 \le x \le 100$$
$$w_2(x) = 100 - .5x, \qquad 0 \le x \le 200.$$

The scattering of disks within $\mathcal{S}$ is according to a standard Poisson process with $\lambda = .01[1/\text{m}^2]$, and the distribution of radii of disks is uniform on (.5,1.0)[m]. Four target points are located on the "other side" of $\mathcal{S}$ with orientation angle

45°, 25°, −10° and −25°. Compute the simultaneous visibility of these four targets.

[3.2.1] Use program VPANN to compute the simultaneous visibility probability, when the scattering region $\mathcal{S}$ is annular, with $u = 50$, $w = 100$[m], $a = .5$, $b = 1.5$[m] and the target points have orientation angles 50°, 45°, 40°, 35°, 30°, 25°, 20°, 15°, 10°, 5° for $\lambda = .001[1/\text{m}^2]$ and for $\lambda = .01[1/\text{m}^2]$.

[3.2.2] A scattering region $\mathcal{S}$ consists of three subregions $\mathcal{S}_1$, $\mathcal{S}_2$, and $\mathcal{S}_3$. $\mathcal{S}_1$ is a horizontal strip with boundaries at $u_1 = 50$[m] and $w_1 = 75$[m]. $\mathcal{S}_2$ is a rectangular section, with lower boundary at $u_2 = w_1 = 75$[m], upper boundary at $w_2 = 100$[m], left boundary at -25[m] and right boundary at 25[m]. $\mathcal{S}_3$ is a horizontal strip with boundaries at $u_3 = w_2 = 100$[m], $w_3 = 125$[m]. Within each strip the scattering is according to a standard Poisson field with $\lambda_1 = .001[1/\text{m}^2]$, $\lambda_2 = .005[1/\text{m}^2]$ and $\lambda_3 = .01[1/\text{m}^2]$. The radii of trees are uniformly distributed between (.5,1.0)[m]. Compute the simultaneous visibility of the four target points, with orientation angles 50°, 25°, 0° and −10°.

[3.2.3] Consider the triangle with vertices $\mathbf{P}_0 = (10, 10)$, $\mathbf{P}_1 = (30, 50)$ and $\mathbf{P}_2 = (5, 50)$. Compute the bivariate normal triangular probability $T(10, 10, 30, 50, 5, 50)$, when the distribution parameters are $\xi = 15$, $\eta = 35$, $\sigma_x = 20$, $\sigma_y = 20$, $\rho = 0.5$.

[3.3.1] Use program VPNORM (see programs for Chapter 5) to evaluate the simultaneous visibility probabilities of $k = 5$ target points having orientation angles 50°, 40°, 30°, 20°, 10°, when the region $\mathcal{S}$ is a horizontal strip with $u = 25$[m], $w = 50$[m], and the scattering of disks is according to a non-homogeneous Poisson field, with *two* clusterings of disks. One according to the bivariate normal distribution with parameters ($\xi = 15$, $\eta = 35$, $\sigma_x = 20$, $\sigma_y = 20$, $\rho = .5$), and the second one according to a bivariate normal with ($\xi = -15$, $\eta = 35$, $\sigma_x = 20$, $\sigma_y = 20$, $\rho = -.5$). The radii of disks are independent of their location, with uniform distribution over (1.0,2.0). The field intensity is $\lambda^* = 10$ [over C^*].

7.3.2. Solutions of Problems For Chapter 3

[3.1.1] The expected value of Y is $\mu_1 = 1.5 \times E\{\text{Beta}(2,2)\} = 1.5/2 = 0.75$[m]. The second moment of Y is $\mu_2 = 1.5^2 \times \dfrac{6}{20} = 0.675$. $\lambda = 0.01[1/\text{m}^2]$. According to Eq. (3.2) $\eta(T) = 0.01(2 \times 100 \times 0.75 + \pi \times 0.675) = 1.5212$.
(i) The number of disks intersecting $\overline{\mathbf{OT}}$ has a Poisson distribution with mean 1.5212.
(ii) $\Pr\{\mathbf{T} \text{ is visible}\} = e^{-1.5212} = 0.218$.

[3.1.2] The origin $\mathbf{O}$ is not covered if no disk of radius Y is centered within a circle of radius Y centered at $\mathbf{O}$. This probability is $\exp\{-\lambda\pi\mu_2\}$. Hence, the conditional probability of $\{L \geq d\}$, given that $\mathbf{O}$ is not covered, is

$$\begin{aligned}\bar{F}(d) &= \exp\{-2\lambda\mu_1 d - \lambda\pi\mu_2\}/\exp\{-\lambda\pi\mu_2\} \\ &= \exp\{-2\lambda\mu_1 d\}.\end{aligned}$$

That is, the conditional distribution of L, given that $\mathbf{O}$ is not covered, is exponential with mean $1/(2\lambda\mu_1)$. In the present problem $\mu_1 = 1.5$ and $2\lambda\mu_1 = 0.3[1/\text{m}]$. Hence, $\bar{F}(30) = \exp\{-9\} = 0.0001$.

[3.1.3] The first two moments of the distribution of Y are $\mu_1 = 0.2$ and $\mu_2 = 0.0433$. According to Eq. (3.5), the distance d[m] from the target should satisfy the inequality

$$\exp\left\{-\lambda\pi\mu_2 - \frac{\lambda}{2}d - \lambda\left(1 + 2\sqrt{d^2 + \frac{1}{4}}\right)\mu_1\right\} \geq 0.8.$$

Substituting $\lambda = 0.02$ and the values of μ_1 and μ_2 we obtain that d should satisfy the inequality

$$0.01d + 0.008\sqrt{d^2 + .25} \leq .2164209.$$

Or

$$(0.008)^2(d^2 + .25) \leq (.2164209 - 0.01d)^2.$$

Expanding the quadratic forms and solving, we obtain $d \leq 12.01$[m].

[3.1.4] The first two moments of the distribution of Y are

$$\mu_1 = 0.648 \int_{0.75}^{1.5} y \exp\left\{-\frac{y - 0.75}{2.5}\right\} dy = 0.464498$$

and

$$\mu_2 = 0.648 \int_{0.75}^{1.50} y^2 \exp\left\{-\frac{y - 0.75}{2.5}\right\} dy = 0.5334572.$$

The visibility probabilities are computed according to Eqs. (3.11)-(3.12).
(i) For $\mathbf{T}_1$ and $\mathbf{T}_4$

r_i	s_i	ϕ
75	15°	0.349066 [rad]
100	−25°	

$$\begin{aligned} E\{A_2(Y;\mathbf{r}^{(2)}, \phi\} &= 350 \times 0.464498 \\ &\quad + \left(\frac{\pi}{2} - 0.349066 - \frac{1}{\tan(0.349066)}\right) \times 0.5334572 \\ &= 161.7603799. \end{aligned}$$

$$\psi_2 = \exp\{-161.7604/100\} = 0.1984.$$

(ii) For $\mathbf{T}_1$, $\mathbf{T}_3$ and $\mathbf{T}_4$ we obtain $\psi_3 = 0.1488$.
(iii) For $\mathbf{T}_1$, $\mathbf{T}_2$, $\mathbf{T}_3$ and $\mathbf{T}_4$ we obtain $\psi_4 = 0.1016$.

[3.1.5] With the new parameters we obtain from Eq. (3.16) the following visibility probabilities, as functions of the distance ρ from **O**

ρ[m]	$\psi_L(s)$
30	0.8114
31	0.6352
32	0.4896
33	0.3925
34	0.3096
35	0.2447
36	0.1939
37	0.1538
38	0.1223
39	0.0973

The distance of **T** from **O** should be 39 [m].

[3.1.6] For the given parameters we obtain

u	25	50	75
ψ_3	0.5148	0.6423	0.8014

[3.1.7] We notice first that the ray from **O** at $s = 5°$ cuts the horizontal line of distance $w = 150$[m] from **O** at $x = 13.12$ [m]. Thus, the lines of sight with orientations $\pm 5°$ are within the extended forest. The visibilities of the targets at $s = \pm 45°$ are independent of the other targets, due to the big distance between the points (verify this!). The simultaneous visibility of the targets at $s = \pm 5°$ is $\psi_2 = 0.0492$. The visibility probability of the target at $\pm 45°$ is $\psi_1 = 0.3462$. Finally, the simultaneous visibility probability of all the four target points is $\psi_4 = 0.0492 \times (0.3462)^2 = 0.0059$.

[3.1.8] Notice that the lines of sights with orientations 45° and 25° pass through $\mathcal{S}_2$, and those with orientations $-10°$ and $-25°$ pass through $\mathcal{S}_1$. Let $\psi_2^{(1)}$ denote the simultaneous visibility probability of the first two targets (45°,25°) and $\psi_2^{(2)}$ that of the last two targets $(-10°, -25°)$. We find $\psi_2^{(1)}$ and $\psi_2^{(2)}$. The joint visibility probability of all four targets ψ_4, is approximately $\psi_2^{(1)} \cdot \psi_2^{(2)}$. The distance of $u_1(x)$ from **O** is $u^{(1)} = 50 \cdot \cos(\alpha)$ where $\alpha = \tan^{-1}(.25)$, i.e., $u^{(1)} = 48.507$. The distance of $w_1(x)$ from **O** is $w^{(1)} = 100\cos(\alpha) = 97.014$. To find $\psi_2^{(1)}$ we can use program SIMVP with the parameters $u^{(1)}$, $w^{(1)}$ and $s_1^* = -10° + (\tan^{-1}(.25))° = 4.036°$ and $s_2^* = -25° + (\tan^{-1}(.25))° = -10.964°$. These parameters yield $\psi_2^{(1)} = 0.2298$. The distance of $u_2(x)$ from **O** is $u^{(2)} = 44.721$ and that of $w_2(x)$ from **O** is $w^{(2)} = 89.443$. The transformed orientation angles are $s_1^* = 45° - (\tan^{-1}(.5))° = 18.435°$ and $s_2^* = 25° - (\tan^{-1}(.5))° = -1.565°$. These parameters yield $\psi_2^{(2)} = .2520$. Finally, $\psi_4 \cong 0.058$.

[3.2.1] The joint visibility probabilities of the 10 points are

λ	0.001	0.01
ψ_{10}	0.3634	0.00

[3.2.2] We notice first that the lines of sight at 20° and 50° do not pass through $\mathcal{S}_2$. Indeed $\tan^{-1}(\frac{25}{75}) = 18.4°$. Let $\psi_4^{(1)}$ be the probability that the four lines of sight are not intersected in $\mathcal{S}_1$. Let $\psi_2^{(2)}$ be the probability that the lines of sight with orientations 0° and −10° are not intersected in $\mathcal{S}_2$. Let $\psi_4^{(3)}$ be the probability that the four lines of sight are not intersected in $\mathcal{S}_3$. The simultaneous visibility probability is (using SIMVP)

$$\psi_4 = \psi_4^{(1)}\psi_2^{(2)}\psi_4^{(3)} = 0.8404 \times 0.6853 \times 0.1758 = 0.101.$$

[3.2.3] The equation of the line connecting $\mathbf{P}_0$ and $\mathbf{P}_1$ is $y = 10 + 2(x - 10)$. The line connecting $\mathbf{P}_0$ and $\mathbf{P}_2$ is $y = 10 - 8(x - 10)$. The triangular probability is

$$\begin{aligned} T(10, 10, 30, 50, 5, 50) = & \frac{1}{20}\int_5^{10} \phi\left(\frac{x}{20}\right)\left[\Phi\left(\frac{50 - 35 - 0.5(x - 15)}{10\sqrt{3}}\right)\right. \\ & - \left.\Phi\left(\frac{10 - 8(x - 10) - 35 - 0.5(x - 15)}{10\sqrt{3}}\right)\right] dx \\ & + \frac{1}{20}\int_{10}^{30} \phi\left(\frac{x}{20}\right)\left[\Phi\left(\frac{50 - 35 - 0.5(x - 15)}{10\sqrt{3}}\right)\right. \\ & - \left.\Phi\left(\frac{10 + 2(x - 10) - 35 - 0.5(x - 15)}{10\sqrt{3}}\right)\right] dx \\ = & \ 0.1492. \end{aligned}$$

[3.3.1] We have to use program VPNORM twice. Once for the first bivariate normal cluster, and the second time for the second. The simultaneous visibility probability of the 5 targets is the product of the values we obtain in these two runs, which is:

$$\psi_5 = 0.24817 \times 0.738913 = .1834.$$

7.4.1. Problems For Chapter 4

[4.1.1] Consider $n = 3$ observation points and $m = 4$ target points. The coordinates (in a scale of 1[m]) of the target points are:

x_i^t	-32	-12	-1	15
y_i^t	100	150	100	120

The coordinates of the observation points are

x_i^0	-10	0	10
y_i^0	0	0	5

a) Find the coefficients α_n, β_n, γ_n of the $N = 12$ lines of sight.
b) Suppose that the centers of random disks are in a strip $\mathcal{S}$, with boundaries at $u = 25$[m] and $w = 60$[m]. Determine the y-coordinates of the points in $\mathcal{S}$ at which the lines of sight intersect.
c) How many substrips should be considered.
d) Compute the order indices of the lines of sight within each substrip. Use Program MOMTVPB with output at MOMTVPB.DAT.

[4.2.1] Consider the target points and observation points of Problem [4.1.1]. The random field is a standard Poisson, and the radii of random disk are uniformly distributed on (0,1.5)[m]. The intensity of the field is $\lambda = 0.002[1/\text{m}^2]$. Use program MOMTVPW to compute the probability that the four target points are simultaneously visible from at *least* two observation points.

[4.2.2] There are $m = 3$ targets and $n = 3$ observation points. The target points are at coordinates (in [m]) (-5,200), (15,200) and (35,200). The coordinates of the observation points are (-10,0), (0,0) and (10,0). The random disks are centered at two parallel strips $\mathcal{S}^{(1)}$ and $\mathcal{S}^{(2)}$. The boundaries of $\mathcal{S}^{(1)}$ are at distances $u^{(1)} = 50$ and $w^{(1)} = 75$. The boundaries of $\mathcal{S}^{(2)}$ are at distances $u^{(2)} = 100$, $w^{(2)} = 125$. The intensity in $\mathcal{S}^{(1)}$ is $\lambda_1 = 0.001[1/\text{m}^2]$ and that in $\mathcal{S}^{(2)}$ is $\lambda_2 = 0.002[1/\text{m}^2]$. The radii of disks are distributed uniformly at (0,0.75). What is the probability that the three targets are simultaneously visible from all the observation points?

[4.2.3] Consider the previous problem. Compute the probabilities of the following events.
(i) Target points $\mathbf{T}_1$ and $\mathbf{T}_3$ are simultaneously visible from $\mathbf{O}_1$ and $\mathbf{O}_2$.
(ii) Target points $\mathbf{T}_2$ and $\mathbf{T}_3$ are simultaneously visible from $\mathbf{O}_1$ and $\mathbf{O}_3$.
(iii) Target points $\mathbf{T}_2$ and $\mathbf{T}_3$ are simultaneously visible from $\mathbf{O}_1$ and $\mathbf{O}_3$ but $\mathbf{T}_1$ is not.
(iv) All three points are *not* simultaneously visible from all the three observation points.

[4.3.1] Derive formulae for the functions $m_k(n, n')$, when the density of disk centers is

$$h(x,y) = \frac{1}{U_k - U_{k-1}} \cdot \frac{1}{2} e^{-|x|},$$

for $-\infty < x < \infty$ and $U_{k-1} \le y \le U_k$. Moreover, the distribution of disk radius is exponential with mean b.

[4.4.1] Redo Problem [4.2.1] with the requirement that around each target point a window of length $L = 2$[m] will be visible.

[4.4.2] Redo Problem [4.2.3] with the requirement that windows of length $L = 1$[m] will be visible around the target points.

[4.5.1] Consider Example 4.4. What is the probability that the window (sphere) of radius 1[m] will be observable around the target point? [Hint: use program THRDVPW. For this you have to compute the inclination angle ϕ. You can assume that the ξ-coordinate of $\mathbf{T}$ is 0.]

[4.5.2] Crowns of trees in a forest are modeled as random spheres, randomly distributed with centers in a layer between $u^* = 20$[m] and $w^* = 40$[m]. A helicopter is flying at an elevation of $r^* = 50$[m] above the ground. The radii of random disks are uniformly distributed with $b = 5$[m]. In order to detect the helicopter from **O**, a window of size greater than 1[m] of the target should be observable. Suppose that the intensity of the random field is $\lambda = 0.001[1/\text{m}^3]$. Show that in order that the detection probability should be smaller than 0.1 the horizontal distance of the helicopter from **O** should be approximately 200[m].

[4.5.3] Redo Example 4.5, with $m = 6$ target points, at coordinates -20, -15, -5, 5, 10, 15 [m].

7.4.2. Solutions of Problems For Chapter 4

[4.1.1] Using program MOMTVPB we obtain the following results:

a)

α_n	β_n	γ_n
−10.000	−0.220	1.0239
−10.000	−0.013	1.0001
−10.000	0.090	1.0040
−10.000	0.208	1.0215
0.000	−0.320	1.0500
0.000	−0.080	1.0032
0.000	−0.010	1.0000
0.000	0.125	1.0078
12.211	−0.442	1.0934
10.759	−0.152	1.0114
10.579	−0.116	1.0067
9.783	0.043	1.0009

b-c) The number of substrips is 13. The values of $U(k)$ are:

1	28.258
2	32.609
3	33.721
4	34.147
5	34.682
6	38.879
7	41.741
8	43.934
9	45.802
10	51.800
11	57.654
12	58.824

d) The ordered line indices in the 13 substrips are:

						Substrips						
1	2	3	4	5	6	7	8	9	10	11	12	13
1	1	1	1	1	1	1	1	1	1	1	1	1
2	2	5	5	5	5	5	5	5	5	5	5	5
5	5	2	2	2	2	2	2	2	2	9	9	9
3	3	3	3	3	3	3	9	9	9	2	2	2
4	4	4	4	9	9	9	3	3	3	3	3	6
6	6	6	9	4	6	6	6	6	6	6	6	3
7	9	9	6	6	4	4	4	4	7	7	7	7
9	7	7	7	7	7	7	7	7	4	4	10	10
8	8	8	8	8	8	10	10	10	10	10	4	4
10	10	10	10	10	10	8	8	11	11	11	11	11
11	11	11	11	11	11	11	11	8	8	8	8	8
12	12	12	12	12	12	12	12	12	12	12	12	12

[4.2.1] The probability that all targets are seen from all 3 observation points is $\psi_{1,2,3} = 0.35416$. The probability that all targets are seen from $\mathbf{O}_1$ and $\mathbf{O}_2$ is $\psi_{1,2} = 0.47142$. The probability that all targets are seen from $\mathbf{O}_1$ and $\mathbf{O}_3$ is $\psi_{1,3} = 0.46308$. The probability that all targets are seen from $\mathbf{O}_2$ and $\mathbf{O}_3$ is $\psi_{2,3} = 0.47143$. The probability that all targets are seen from **at least** two observation points is $\psi_{1,2} + \psi_{1,3} + \psi_{2,3} - 2 \cdot \psi_{1,2,3} = 0.6976$.

[4.2.2] The simultaneous visibility probability through $\mathcal{S}_1$ is $\psi^{(1)} = 0.86098$, that through $\mathcal{S}_2$ is $\psi^{(2)} = 0.71734$. The simultaneous visibility through $\mathcal{S}_1$ and $\mathcal{S}_2$ is $\psi = \psi^{(1)} \times \psi^{(2)} = 0.6176$.

[4.2.3] Let ψ_{i,i_2-j_1,j_2} denote the simultaneous probability of observing $\mathbf{T}_{i_1}$ and $\mathbf{T}_{i_2}$ from $\mathbf{O}_{j_1}$ and $\mathbf{O}_{j_2}$.

(i) $\psi_{1,3-1,2} = 0.7967$

(ii) $\psi_{2,3-1,3} = 0.7997$

(iii) The probability of observing all three targets from $\mathbf{O}_1$ and $\mathbf{O}_3$ is $\psi_{1,2,3-1,3} = 0.71909$. The answer is $\psi_{2,3-1,3} - \psi_{1,2,3-1,3} = 0.0806$.

(iv) $1 - 0.6176 = 0.3824$ [see answer to previous Problem].

[4.3.1] According to Eq. (4.23) the formula for $m_k(n, n')$, $1 \leq n < n' \leq N$ is

$$m_k(n,n') = \frac{\lambda\gamma_n}{2(U_k - U_{k-1})b} \int_{U_{k-1}}^{U_k} \int_0^{\xi_{nn'}(y)} \exp\left\{-\frac{\rho}{b} - |\gamma_n\rho + x_n(y)|\right\} d\rho dy$$
$$+ \frac{\lambda\gamma_{n'}}{2(U_k - U_{k-1})b} \int_{U_{k-1}}^{U_k} \int_0^{\xi_{nn'}(y)} \exp\left\{-\frac{\rho}{b} - |-\gamma_{n'}\rho + x_{n'}(y)|\right\} d\rho dy.$$

[4.4.1] For $L = 2$, the probability that all the four targets are visible from **all** three observation points is $\psi(L) = 0.24766$. The corresponding probabilities from observation points $\mathbf{O}_{j_1}$, $\mathbf{O}_{j_2}$ is $\psi_{j_1,j_2}(L)$. We obtain

$$\psi_{1,2}(2) = 0.33875,$$
$$\psi_{1,3}(2) = 0.34621,$$
$$\psi_{2,3}(2) = 0.34349.$$

The probability that the four targets will be visible from **at least** two observation points is 0.5331.

[4.4.2] For $L = 1$,
(i) $\psi_{1,3-1,2}(L) = 0.89851 \times 0.76795 = 0.6900$
(ii) $\psi_{2,3-1,3}(1) = 0.89851 \times 0.77509 = 0.6964$
(iii) $\psi_{1,2,3-1,3}(1) = 0.85595 \times 0.68594 = 0.5871$. Answer: $\psi_{2,3-1,3}(1) - \psi_{1,2,3-1,3}($
0.1029.

[4.5.1] The inclination angle is $\tan^{-1}\left(\frac{200}{1000}\right) = 11.31°$. Program THRDVPW yields the visibility probability $\psi = 0.6325$.

[4.5.2] Using program THRDVPW we find that the visibility probability with inclination angle $\phi = 75°$ is 0.104 and with $\phi = 76°$ is 0.0896. Thus the distance from $\mathbf{O}$ should be $d = 50 \tan(75.5°) = 193.3$ [m].

[4.5.3] For the 6 targets we obtain the following visibility probabilities

L	b	$\phi = 15°$	$\phi = 45°$	$\phi = 60°$
0	1	0.8489	0.8001	0.7299
0	3	0.2575	0.1575	0.0735
1	1	0.7194	0.6865	0.5247
1	3	0.1739	0.0920	0.0343

7.5.1. Problems for Chapter 5

[5.1.1] Consider $m = 4$ targets and $\nu = 2$ observation points. There are $N = 8$ possible lines of sight. Express the probabilities of the following elementary visibility events in terms of visibility probabilities
(i) Only lines $\mathcal{L}_1$ and $\mathcal{L}_8$ are visible.
(ii) Only line $\mathcal{L}_7$ is not visible.
(iii) Lines $\mathcal{L}_2$, $\mathcal{L}_4$, $\mathcal{L}_6$ and $\mathcal{L}_8$ are visible and the other lines are invisible.

[5.1.2] Redo Example 5.4 with a standard Poisson field of intensity of $\lambda = 0.002$ [1/m^2] and radii of disk uniformly distributed over $(0, 2)$ [m].

[5.1.3] Write the matrix of coefficients H_4, and the 16 probabilities π_l ($l = 0, \cdots, 2^4 - 1$ in terms of the visibility probabilities $\psi\{\cdot\}$.

[5.1.4] Use program VPWALL to determine the visibility probabilities in the case of $m = 4$ targets and $\nu = 1$ observation points. The random field is standard Poisson. The strip $\mathcal{S}$ is between two parallel lines. $u = 35$ [m], $w = 115$ [m $\lambda = 0.001$ [1/m^2], the radii of disks are uniformly distributed on $(0, 0.5)$ [m The coordinates of the targets are (in meters) $(-25, 100)$, $(-15, 110)$, $(0, 12($ and $(15, 100)$. The observation point is at $(0, 0)$. The required window size is $L = 1$ [m]. What is the probability that only targets 1 and 4 are visible

Compute the probability distribution of the number of targets, N, observable from $\mathbf{0}$.

[5.2.1] Consider a curve $\mathcal{C}$, whose rectangular coordinates presentation is

$$y(x) = r + h - |x|\frac{h}{x_U}, \qquad \text{for} \quad -x_U \le x \le x_U$$

where r and h are positive parameters, $x_U = r\,\tan(\theta_U)$. For orientation angles s, $-\theta_U < s < \theta_U$, show that $\mathcal{C}$ can be represented by the function

$$\rho(s) = \frac{r(1+\tan^2(s))^{1/2}}{m(s)}\tan(\theta_U),$$

where $m(s) = \dfrac{r\,\tan(\theta_U) + h\tan(s)}{r+h}$.

[5.2.2] If V is the visibility measure given in Eq. (5.15) and L is the length of the rectifiable curve, $\bar{\mathcal{C}}$, from θ_L to θ_U, then

$$\lim_{n\to\infty} E\{V^n\} = \begin{cases} \infty, & \text{if } L > 1 \\ 0, & \text{if } L < 1. \end{cases}$$

[5.3.1] The first two moments of W can be approximated numerically, in the circular case (see Example 5.7) by the following method. Partition the interval (θ_L, θ_U) to np equal subinterval. Let $s_i = \theta_U - \left(i - \frac{1}{2}\right)\delta$ where $\delta = (\theta_U - \theta_L)/np$. Then

$$E\{W\} \cong \frac{1}{np}\sum_{i=1}^{np}\psi_1(s_i).$$

Similarly, for $k \ge i$, let

$$s_k = \theta_U - k\delta$$

then

$$E\{W^2\} \cong \frac{1}{np(np+1)}\sum_{i=1}^{np}\sum_{j=i}^{np}\psi_2(s_i, s_k).$$

Program VPANNIN computes the first two moments of W in this manner. Compute the mean and standard deviation of W, in the circular case, when $u = 30$ [m], $w = 55$ [m], $a = 1$ [m], $b = 3.5$ [m], $\theta_L = -\pi/6$, $\theta_U = \pi/6$ and $\lambda = 0.0002$ [1/m^2].

[5.4.1] Use program MOANNTSM to estimate the moments and the parameters of the mixed-beta approximation to the distribution of W in the annular case, when the field parameters are: $u = 30$ [m], $w = 50$ [m], $r = 100$ [m], $a = 1$ [m], $b = 3$ [m], $\theta_L = 0$, $\theta_U = \pi/2$, and $\lambda = 0.003$ [1/m^2]. With these parameters, estimate

(i) $\Pr\{W \le 0.5\}$.

(ii) The quartiles and the median of W.

7.5.2. Solutions of Problems For Chapter 5

[5.1.1] (i) By expanding $O_1(1-O_2)(1-O_3)(1-O_4)(1-O_5)(1-O_6)(1-O_7)O_8$ we obtain

$$\begin{aligned}
&\Pr\{\text{Only } \mathcal{L}_1 \text{ and } \mathcal{L}_8 \text{ are visible}\} = \psi(1,8)\\
&-\psi(1,2,8)-\psi(1,3,8)-\psi(1,4,8)-\psi(1,5,8)-\psi(1,6,8)\\
&-\psi(1,7,8)+\psi(1,2,3,8)+\psi(1,2,4,8)+\psi(1,2,5,8)\\
&+\psi(1,2,6,8)+\psi(1,2,7,8)+\psi(1,3,4,8)+\psi(1,3,5,8)\\
&+\psi(1,3,6,8)+\psi(1,3,7,8)+\psi(1,4,5,8)+\psi(1,4,6,8)\\
&+\psi(1,4,7,8)+\psi(1,5,6,8)+\psi(1,5,7,8)+\psi(1,6,7,8)\\
&-\psi(1,2,3,4,8)-\psi(1,2,3,5,8)-\psi(1,2,3,6,8)\\
&-\psi(1,2,3,7,8)-\psi(1,2,4,5,8)-\psi(1,2,4,6,8)\\
&-\psi(1,2,4,7,8)-\psi(1,2,5,6,8)-\psi(1,2,5,7,8)\\
&-\psi(1,2,6,7,8)-\psi(1,3,4,5,8)-\psi(1,3,4,6,8)\\
&-\psi(1,3,4,7,8)-\psi(1,3,5,6,8)-\psi(1,3,5,7,8)\\
&-\psi(1,3,6,7,8)-\psi(1,4,5,6,8)-\psi(1,4,5,7,8)\\
&-\psi(1,4,6,7,8)-\psi(1,5,6,7,8)+\psi(1,2,3,4,5,8)\\
&+\psi(1,2,3,4,6,8)+\psi(1,2,3,4,7,8)+\psi(1,2,3,5,6,8)\\
&+\psi(1,2,3,5,7,8)+\psi(1,2,3,6,7,8)+\psi(1,2,4,5,6,8)\\
&+\psi(1,2,4,5,7,8)+\psi(1,2,4,6,7,8)+\psi(1,2,5,6,7,8)\\
&+\psi(1,3,4,5,6,8)+\psi(1,3,4,5,7,8)+\psi(1,3,4,6,7,8)\\
&+\psi(1,3,5,6,7,8)+\psi(1,4,5,6,7,8)-\psi(1,2,3,4,5,6,8)\\
&-\psi(1,2,3,4,5,7,8)-\psi(1,2,3,4,6,7,8)\\
&-\psi(1,2,3,5,6,7,8)-\psi(1,2,4,5,6,7,8)\\
&-\psi(1,3,4,5,6,7,8)+\psi(1,2,3,4,5,6,7,8).
\end{aligned}$$

(ii) $\Pr\{\text{Only } \mathcal{L}_7 \text{ is not visible}\} = \psi(1,2,3,4,5,6,8)-\psi(1,2,3,4,5,6,7,8)$.

(iii)

$$\begin{aligned}
&\Pr\{\text{Only } \mathcal{L}_2,\mathcal{L}_4,\mathcal{L}_6 \text{ and } \mathcal{L}_8 \text{ are visible}\}\\
&=\psi(2,4,6,8)-\psi(1,2,4,6,8)-\psi(2,3,4,6,8)\\
&-\psi(2,4,5,6,8)-\psi(2,4,6,7,8)+\psi(1,2,3,4,6,8)\\
&+\psi(1,2,4,5,6,8)+\psi(1,2,4,6,7,8)\\
&+\psi(2,3,4,5,6,8)+\psi(2,3,4,6,7,8)\\
&+\psi(2,4,5,6,7,8)-\psi(1,2,3,4,5,6,8)\\
&-\psi(1,2,3,4,6,7,8)-\psi(1,2,4,5,6,7,8)\\
&-\psi(2,3,4,5,6,7,8)+\psi(1,2,3,4,5,6,7,8).
\end{aligned}$$

[5.1.2] In Example 5.4 the 5 target points are specified along a circular curve. We take here the targets on a line parallel to W at distance $r = 100$ [m]. The x-coordinates of the targets are given by $x = 100 \times \tan(\theta)$. These are: 46.631, 26.795, 8.749, −8.749 and −26.795. Using program VPWALL with $u = 20$ [m], $w = 60$ [m], $\lambda = 0.002$ [1/m^2] and $b = 2$ [m] we obtain the following visibility probabilities

i	j	k	l	m	ψ	i	j	k	l	m	ψ
1					.8474					5	.8382
	2				.8516	1				5	.7102
1	2				.7217		2			5	.7138
		3			.8516	1	2			5	.6049
1	3				.7216			3		5	.7138
	2	3			.7253	1		3		5	.6048
1	2	3			.6146		2	3		5	.6079
			4		.8474	1	2	3		5	.5152
1			4		.7180				4	5	.7102
	2		4		.7216	1			4	5	.6018
1	2		4		.6115		2		4	5	.6048
		3	4		.7217	1	2		4	5	.5215
1		3	4		.6115			3	4	5	.6049
	2	3	4		.6146	1		3	4	5	.5215
1	2	3	4		.5208	1	2	3	4	5	.4365

The corresponding elementary probabilities are:

ν	i_1	i_2	i_3	i_4	i_5	ψ	ν	i_1	i_2	i_3	i_4	i_5	ψ
31	1	1	1	1	1	0.4365	24	−1	−1	−1	1	1	0.0024
30	−1	1	1	1	1	0.0786	20	−1	−1	1	−1	1	0.0025
29	1	−1	1	1	1	0.0760	12	−1	−1	1	1	−1	0.0027
27	1	1	−1	1	1	0.0760	18	−1	1	−1	−1	1	0.0025
23	1	1	1	−1	1	0.0786	10	−1	1	−1	1	−1	0.0027
15	1	1	1	1	−1	0.0843	6	−1	1	1	−1	−1	0.0027
28	−1	−1	1	1	1	0.0137	17	1	−1	−1	−1	1	0.0024
26	−1	1	−1	1	1	0.0137	9	1	−1	−1	1	−1	0.0026
22	−1	1	1	−1	1	0.0142	5	1	−1	1	−1	−1	0.0026
14	−1	1	1	1	−1	0.0152	3	1	1	−1	−1	−1	0.0026
25	1	−1	−1	1	1	0.0133	16	−1	−1	−1	−1	1	0.0004
21	1	−1	1	−1	1	0.0137	8	−1	−1	−1	1	−1	0.0004
13	1	−1	1	1	−1	0.0147	4	−1	−1	1	−1	−1	0.0004
19	1	1	−1	−1	1	0.0137	2	−1	1	−1	−1	−1	0.0004
11	1	1	−1	1	−1	0.0147	1	1	−1	−1	−1	−1	0.0005
7	1	1	1	−1	−1	0.0152	0	−1	−1	−1	−1	−1	0.0001

$i_j = -1$ $(j = 1, \cdots, 5)$ in the above table signifies that the j-th target is invisible. ν is the standard order index, i.e., $\nu = \sum_{j=1}^{5} b_j 2^{j-1}$, where

$$b_j = \begin{cases} 1, & \text{if } i_j = 1 \\ \\ 0, & \text{if } i_j = -1. \end{cases}$$

The distribution of the number of visible targets is:

j	0	1	2	3	4	5
$\Pr\{J = j\}$	0.0001	0.0021	0.0257	0.1421	0.3935	0.4365

Here also the median of J is $Me = 4$ and the expected number is $E\{J\} = 4.2363$.

[5.1.3]

$$H_4 = \begin{bmatrix}
1 & -1 & -1 & 1 & -1 & 1 & 1 & -1 & -1 & 1 & 1 & -1 & 1 & -1 & -1 & 1 \\
0 & 1 & 0 & -1 & 0 & -1 & 0 & 1 & 0 & -1 & 0 & 1 & 0 & 1 & 0 & -1 \\
0 & 0 & 1 & -1 & 0 & 0 & -1 & 1 & 0 & 0 & -1 & 1 & 0 & 0 & 1 & -1 \\
0 & 0 & 0 & 1 & 0 & 0 & 0 & -1 & 0 & 0 & 0 & -1 & 0 & 0 & 0 & 1 \\
0 & 0 & 0 & 0 & 1 & -1 & -1 & 1 & 1 & -1 & -1 & 1 & -1 & 1 & 1 & -1 \\
0 & 0 & 0 & 0 & 0 & 1 & 0 & -1 & 0 & 1 & 0 & -1 & 0 & -1 & 0 & 1 \\
0 & 0 & 0 & 0 & 0 & 0 & 1 & -1 & 0 & 0 & 1 & -1 & 0 & 0 & -1 & 1 \\
0 & 0 & 0 & 0 & 0 & 0 & 0 & 1 & 0 & 0 & 0 & 1 & 0 & 0 & 0 & -1 \\
 & & & & & & & & 1 & -1 & -1 & 1 & -1 & 1 & 1 & -1 \\
 & & & & & & & & 0 & 1 & 0 & -1 & 0 & -1 & 0 & 1 \\
 & & & & & & & & 0 & 0 & 1 & -1 & 0 & 0 & -1 & 1 \\
 & & & & & & & & 0 & 0 & 0 & 1 & 0 & 0 & 0 & -1 \\
 & & & 0 & & & & & 0 & 0 & 0 & 0 & 1 & -1 & -1 & 1 \\
 & & & & & & & & 0 & 0 & 0 & 0 & 0 & 1 & 0 & -1 \\
 & & & & & & & & 0 & 0 & 0 & 0 & 0 & 0 & 1 & -1 \\
 & & & & & & & & 0 & 0 & 0 & 0 & 0 & 0 & 0 & 1
\end{bmatrix}$$

$$\begin{aligned}
\pi_0 &= 1 - \psi(1) - \psi(2) + \psi(1,2) - \psi(3) + \psi(1,3) + \psi(2,3) - \psi(1,2,3) \\
&\quad - \psi(4) + \psi(1,4) + \psi(2,4) - \psi(1,2,4) + \psi(3,4) - \psi(1,3,4) \\
&\quad - \psi(2,3,4) + \psi(1,2,3,4) \\
\pi_1 &= \psi(1) - \psi(1,2) - \psi(1,3) + \psi(1,2,3) - \psi(1,4) + \psi(1,2,4) \\
&\quad + \psi(1,3,4) - \psi(1,2,3,4) \\
\pi_2 &= \psi(2) - \psi(1,2) - \psi(2,3) + \psi(1,2,3) - \psi(2,4) + \psi(1,2,4) \\
&\quad + \psi(2,3,4) - \psi(1,2,3,4) \\
\pi_3 &= \psi(1,2) - \psi(1,2,3) - \psi(1,2,4) + \psi(1,2,3,4) \\
\pi_4 &= \psi(3) - \psi(1,3) - \psi(2,3) + \psi(1,2,3) + \psi(4) - \psi(1,4) \\
&\quad - \psi(2,4) + \psi(1,2,4) - \psi(3,4) + \psi(1,3,4) + \psi(2,3,4) - \psi(1,2,3,4)
\end{aligned}$$

$$\pi_5 = \psi(1,3) - \psi(1,2,3) + \psi(1,4) - \psi(1,2,4) - \psi(1,3,4) + \psi(1,2,3,4)$$
$$\pi_6 = \psi(2,3) - \psi(1,2,3) + \psi(2,4) - \psi(1,2,4) - \psi(2,3,4) + \psi(1,2,3,4)$$
$$\pi_7 = \psi(1,2,3) + \psi(1,2,4) - \psi(1,2,3,4)$$
$$\pi_8 = \psi(4) - \psi(1,4) - \psi(2,4) + \psi(1,2,4) - \psi(3,4) + \psi(1,3,4) + \psi(2,3,4) - \psi(1,2,3,4)$$
$$\pi_9 = \psi(1,4) - \psi(1,2,4) - \psi(1,3,4) + \psi(1,2,3,4)$$
$$\pi_{10} = \psi(2,4) - \psi(1,2,4) - \psi(2,3,4) + \psi(1,2,3,4)$$
$$\pi_{11} = \psi(1,2,4) - \psi(1,2,3,4)$$
$$\pi_{12} = \psi(3,4) - \psi(1,3,4) - \psi(2,3,4) + \psi(1,2,3,4)$$
$$\pi_{13} = \psi(1,3,4) - \psi(1,2,3,4)$$
$$\pi_{14} = \psi(2,3,4) - \psi(1,2,3,4)$$
$$\pi_{15} = \psi(1,2,3,4)$$

[5.1.4] (i) The probability that only targets 1 and 4 are visible is $\pi_9 = \psi(1,4) - \psi(1,2,4) - \psi(1,3,4) + \psi(1,2,3,4)$ [see previous problem]. According to program VPWALL with the given parameters we get

$$\psi(1,4) = 0.81736$$
$$\psi(1,2,4) = 0.73924$$
$$\psi(1,3,4) = 0.74701$$
$$\psi(1,2,3,4) = 0.67562.$$

Hence

$$\pi_9 = 0.00673.$$

In a similar manner we obtain (one can write a computer program that computes all these elementary probabilities from the output of VPWALL).

ν	π_ν	ν	π_ν
0	0.00010	8	0.00070
1	0.00070	9	0.00673
2	0.00070	10	0.00679
3	0.00673	11	0.06362
4	0.00080	12	0.00761
5	0.00754	13	0.07139
6	0.00761	14	0.07197
7	0.07139	15	0.67562

(ii) The distribution of the number of visible targets, N is

$$\begin{aligned}
\Pr\{N=0\} &= \pi_0 = 0.0001\\
\Pr\{N=1\} &= \pi_1+\pi_2+\pi_4+\pi_8 = 0.0029\\
\Pr\{N=2\} &= \pi_3+\pi_5+\pi_6+\pi_9+\pi_{10}+\pi_{12}\\
&= 0.0430\\
\Pr\{N=3\} &= \pi_7+\pi_{11}+\pi_{13}+\pi_{14} = 0.2784
\end{aligned}$$

and

$$\Pr\{N=4\} = \pi_{15} = 0.6756.$$

[5.2.1] Let $(x(x), y(s))$ be the coordinates of the point on $\mathcal{C}$, at which a ray with orientation s intersects. Without loss of generality, assume that $s \geq 0$. Notice that

$$\begin{aligned}
\rho(s) &= x(s)(1+(\cot(s))^2)^{1/2}\\
&= \frac{x(s)}{\tan(s)}(1+\tan^2(s))^{1/2}.
\end{aligned}$$

Moreover,

$$\begin{aligned}
y(s) &= x(s)\cot(s)\\
&= r+h-x(s)\frac{h}{r\ \tan(\theta_U)}.
\end{aligned}$$

Hence,

$$x(s)\left(\frac{1}{\tan(s)}+\frac{h}{r\ \tan(\theta_U)}\right) = r+h.$$

Thus,

$$x(s) = \frac{r\ \tan(s)\tan(\theta_U)}{\dfrac{r\ \tan(\theta_U)+h\ \tan(s)}{r+h}} = \frac{r\ \tan(s)\tan(\theta_U)}{m(s)}.$$

Substituting above we obtain

$$\rho(s) = \frac{r\ \tan(\theta_U)}{m(s)}(1+\tan^2(s))^{1/2}.$$

[5.2.2] $V^n = W^n \cdot L^n$. Thus, according to Eq. (5.19),

$$\begin{aligned}
E\{V^n\} &= p_1 L^n + L^n \int_0^1 w^n dH_w^*(w)\\
&\underset{n\to\infty}{\longrightarrow} \begin{cases} \infty, & \text{if } L>1\\ 0, & \text{if } L<1. \end{cases}
\end{aligned}$$

[5.3.1] The results for $np = 50$ are $\mu_1 = 0.9822$ and $\mu_2 = 0.9667$.

[5.4.1] With $NR = 200$ replicas we obtain the estimates:

Order	1	2	3	4	5	6
Moment	.7851	.6411	.5422	.4905	.4583	.4409

$$p_0 \cong 0, \quad p_1 \cong 0.0181.$$

The parameters of the Beta distribution are $\hat{\alpha} = 2.9268$, $\hat{\beta} = 0.9074$. The distribution of W is approximated by:

$$H_W(w) \cong 0.9819 I_w(2.9268, 0.9074),$$

where $I_w(\alpha, \beta)$ is the incomplete beta function ratio.

(i)

$$\begin{aligned} \Pr\{W \leq 0.5\} &\cong 0.9819 \times I_{0.5}(2.9268, 0.9074) \\ &= 0.9819 \times 0.1155 \\ &= 0.1134. \end{aligned}$$

(ii)

$$\begin{aligned} Q_1 &= 0.65 \\ Me &= 0.815 \\ Q_3 &= .927. \end{aligned}$$

To obtain these results one needs a program for computing $I_w(\alpha, \beta)$. One can use MATHCAD® or MATHEMATICA®.

7.6.1. Problems for Chapter 6

[6.1.1] Use program VIEWLNG to determine the mean, median and standard deviation of the distribution of the visible segment to the right of $x = 10$ [m], under the field parameters $r = 200$ [m], $u = 100$ [m], $w = 150$ [m], $x_L = -100$ [m], $x_U = 100$ [m], $a = 2$ [m], $b = 3$ [m] and $\lambda = 0.003$ [1/m^2].

[6.1.2] Consider a trapezoidal region with field parameters as in the previous exercise. Compute the probability that the interval on $\bar{C}$, between $x = 50$ [m] and $x = 75$ [m] is completely visible. [Hint: Compute first the visibility probability of $\mathbf{P}_{50}$. Then multiply this by $\bar{V}(25 \mid 50)$].

[6.2.1] The function $K^*(x,t)$ given by Eq. (6.4) was derived for the case where the radius, Y, of an obscuring disk has a uniform distribution on (a, b). Suppose that the p.d.f. of Y on (a, b) is

$$f_Y(y) = \begin{cases} \dfrac{2}{(b-a)^2}(y-a), & \text{if } a < y < b \\ 0, & \text{otherwise.} \end{cases}$$

Derive the function $K^*(x,t)$ from Eq. (6.3), for the new p.d.f. of Y.

[6.3.1] Derive Eqs. (6.5) and (6.6).

[6.3.2] Derive the c.d.f. $H_Z(z)$, given by Eq. (6.8).

[6.3.3] Verify Eq. (6.9).

[6.3.4] Since the function $S(z;x)$, given by Eq. (6.7), is an increasing function of z, the fractiles of $U_s(x)$ can be obtained by substituting the fractiles of Z (obtained from Eq. (6.8)) in Eq. (6.7). What are the median, the 10th and 90th percentiles of $U_s(x)$ when $x = 5$, $r = 150$, $u = 90$, $w = 100$, $a = 2$, $b = 4$?

[6.3.5] Compute $E\{U_s(x)\}$ for the field parameters of the previous exercise and $x = 5$. [Hint: $E\{U_s(x)\} = \int_0^\infty (1 - Q_s(\eta \mid x))d\eta$. Use program SNGSHDW.]

[6.3.6] Show that in the annular case (both $\mathcal{U}$, $\mathcal{W}$ and $\mathcal{C}$ are on concentric circles of radii u, w and r) the c.d.f. of the length L of a shadowed arc on $\mathcal{C}$, cast by a single disk, in the standard-uniform case, is

$$\begin{aligned} G_s(l) = {} & \frac{2\sin(\frac{l}{2r})}{3(b-a)(w^2-u^2)}\left[\left(\min\left(\frac{b}{\sin(\frac{l}{2r})}, w\right)\right)^3 \right. \\ & \left. - \left(\max\left(\frac{a}{\sin(\frac{l}{2r})}, u\right)\right)^3\right] \\ & - \frac{a}{(b-a)(w^2-u^2)}\left[\left(\min\left(\frac{b}{\sin(\frac{l}{2r})}, w\right)\right)^2 \right. \\ & \left. - \left(\max\left(\frac{a}{\sin(\frac{l}{2r})}, u\right)\right)^2\right] \\ & + \frac{1}{w^2-u^2}\left[w^2 - \left(\min\left(\frac{b}{\sin(\frac{l}{2r})}, w\right)\right)^2\right]. \end{aligned}$$

[6.5.1] Consider the trapezoidal case. Let $[x_1, x_2]$ be an interval on $\bar{\mathcal{C}}$, $x_L < x_1 < x_2 < x_U$. We are interested to compute the probability, p_0, that this interval is completely shadowed, given that the shadow starts at x. This probability is

$$p_0 = 1 - D_T(x_2 \mid x_1).$$

Use program CDFSHDW to compute $D_T(x_2 \mid x_1)$ and p_0 when, $x_L = -100$, $x_U = 100$, $x_1 = -15$, $x_2 = 0$, $r = 100$, $u = 40$, $w = 60$, $a = 1$, $b = 2$ and $\lambda = 0.05$ [1/m^2].

[6.5.2] Use program CDFSHDW to compute the mean and standard deviation of the length of a shadow starting at $x = 0$ [m] when the field parameters (in [m]) are $x_L^* = -100$, $x_U^* = 100$, $r = 100$, $u = 30$, $w = 60$, $a = 2$, $b = 4$, $\lambda = 0.001$ [1/m^2].

[6.6.1] Use program DISTNSH to compute the probability that in the interval (0, 30) there will be
(i) at most one shadow;
(ii) at least 3 shadows;
when the field parameters (in [m]) $x_L^* = -100$, $x_U^* = 100$ are $r = 100$, $u = 50$, $w = 75$, $a = 0$, $b = 3$ and $\lambda = 0.02$ [1/m^2].

[6.7.1] Use program SURVFUC to compute the probability that a target will survive moving along a path of 30 [m] stretch, starting at $x = 0$ when $q = -ln(.8)$ and the field parameters (in [m]) are $r = 100$, $u = 50$, $w = 60$, $a = 0$, $b = 3$ and $\lambda = 0.01$. $x_L^* = -100$ and $x_U^* = 100$.

7.6.2. Solutions of Problems For Chapter 6

[6.1.1] For the parameters of the problem, program VIEWLNG yields the following values of $\bar{V}(l \mid 10)$

l	5	10	15	20	25	30
$\bar{V}(l \mid 10)$	0.6254	0.3911	0.2445	0.1528	0.0955	0.0596

l	35	40	45	50	55	60
$\bar{V}(l \mid 10)$	0.0373	0.0233	0.0145	0.0091	0.0057	0.0035

From this value we obtain $M_e \cong 7.675$ (linear interpolation)

$$\begin{aligned}\text{Mean} &= 5 \times \sum_{j=1}^{20} (\bar{V}(5j \mid 10) + \bar{V}(5(j-1) \mid 10))/2 \\ &= 10.8095,\end{aligned}$$

where $\bar{V}(0 \mid 10) = 1$. This formula for approximating the expected value of L, given x, is based on the formula

$$E(L \mid x) = \int_0^\infty \bar{V}(l \mid x) dl.$$

Since we are given only the values of $V(5 \cdot j \mid x)$, we approximate $\int_{5(j-1)}^{5j} \bar{V}(l \mid x) dl$ by the area of the rectangle $5(\bar{V}(5(j-1) \mid x) + \bar{V}(5j \mid x))/2$. Similarly, the second moment of L can be approximated by

$$\begin{aligned}E\{L^2 \mid x\} &= 2\int_0^\infty t\bar{V}(t \mid x) dt \\ &= 25 \sum_{j=1}^{\infty} ((j-1)\bar{V}(5(j-1) \mid x) + j\bar{V}(5j \mid x).\end{aligned}$$

In our case we have $E\{L^2 \mid 10\} \cong 217.3125$. Hence, Standard Deviation $(L \mid 10) \cong 10.0233$.

[6.1.2] The visibility probability of P_{50} is (program VPTRAP) $\psi(50) = 0.46$. Program VIEWLNG yields $\bar{V}(25 \mid 50) = 0.0946$. Thus, the answer is: 0.04352.

[6.2.1] We distinguish between six cases.

Case I: $\dfrac{dw}{t} \le a$,

Case II: $\dfrac{du}{t} < a < \dfrac{dw}{t} \le b$,

Case III: $\dfrac{du}{t} < a < b < \dfrac{dw}{t}$,

Case IV: $a \le \dfrac{du}{t} < \dfrac{dw}{t} \le b$,

Case V: $a \le \dfrac{du}{t} < b < \dfrac{dw}{t}$,

Case VI: $b \le \frac{du}{t}$.
The function $K^*(x,t)$ assumes the following expressions:
Case I: $K^*(x,t) = 0$.
Case II:

$$K^*(x,t) = \frac{1}{tw(b-a)^2} \int_a^{\frac{dw}{t}} (tw - yd)^2 y dy$$
$$= \frac{1}{tw(b-a)^2} \left[\frac{d^2}{2} \left(w^4 - \frac{a^4}{2} \right) + \frac{2}{3} a^3 dtw - \frac{1}{2} a^2 t^2 w^2 - \frac{2}{3} \frac{d^4 w^4}{t^2} + \frac{1}{4} \frac{d^6 w^4}{t^4} \right].$$

Case III:

$$K^*(x,t) = \frac{1}{tw(b-a)^2} \left[\frac{d^2}{4} (b^4 - a^4) - \frac{2}{3} dtw (b^3 - a^3) + \frac{1}{2} t^2 w^2 (b^2 - a^2) \right].$$

Case IV:

$$K^*(x,t) = \frac{1}{tw(b-a)^2} \left[\frac{1}{2} d^2 w^2 (w^2 - u^2) - \frac{2}{3} \frac{wd^4}{t^2} (w^3 - u^3) + \frac{1}{4} \frac{d^6}{t^4} (w^4 - u^4) \right]$$
$$+ \frac{2}{(b-a)^2} \left[\frac{2}{3} \frac{a^3 d}{(u+w)} - \frac{1}{4} \frac{a^2}{w} (w^2 - u^2) t + \frac{1}{4} \frac{d^2 u^2}{w} (w^2 - u^2) \frac{1}{t} - \frac{2}{3} \frac{d^4 u^3}{u+w} \frac{1}{t^3} \right].$$

Case V:

$$K^*(x,t) = \frac{1}{tw(b-a)^2} \left[\frac{d^2}{2} \left(\frac{b^4}{2} - u^2 w^2 \right) - \frac{2}{3} b^3 dwt + \frac{1}{2} b^2 wt^2 + \frac{2}{3} d^4 u^3 w \cdot \frac{1}{t^2} - \frac{1}{4} d^6 u^4 \cdot \frac{1}{t^4} \right]$$
$$+ \frac{2}{(b-a)^2} \left[\frac{2}{3} \cdot \frac{a^3 d}{u+w} - \frac{1}{4} \frac{a^2}{w} (w^2 - u^2) t + \frac{1}{4} \frac{d^2 u^2}{w} (w^2 - u^2) \cdot \frac{1}{t} - \frac{2}{3} \frac{d^4 u^3}{u+w} \cdot \frac{1}{t^3} \right].$$

Case VI:

$$K^*(x,t) = \frac{2}{(b-a)^2} \left[-\frac{2}{3} \cdot \frac{d}{u+w} (b^3 - a^3) + \frac{1}{4} \frac{(b^2 - a^2)}{w} (w^2 - u^2 t \right].$$

[6.3.1] Let $\mathbb{R}_x$ be a ray from **O** which cuts the line C at the point (x, r). The ray $\mathbb{R}$ cuts the line $\mathcal{V}$ at $(x\frac{v}{r}, v)$. The center of a disk tangential to $\mathbb{R}_x$ with radius Y

on $\mathcal{V}$, has coordinates (x_c, v), where $x_c = x\frac{v}{r} + h$. Moreover, if $s = \tan^{-1}(\frac{x}{r})$ is the orientation angle of $\mathbb{R}_x$, then

$$h = \frac{Y}{\cos(s)} = \frac{Y}{r}(x^2 + r^2)^{1/2}$$
$$= Y\left(1 + \left(\frac{x}{r}\right)^2\right)^{1/2}.$$

This proves Eq. (6.5). Let $\mathbb{R}_x^+$ be the ray from **O** tangential to the disk, to the right of x_c. The angle between $\mathbb{R}_x$ and $\mathbb{R}_x^+$ is $\alpha = 2\sin^{-1}(\frac{Y}{\sqrt{x_c^2 + v^2}})$. Hence, $\mathbb{R}_x^+$ cuts $\mathcal{C}$ at the point (U_x, r), where

$$U_x = r\ \tan(s + \alpha) = r\ \tan\left(2\sin^{-1}\left(\frac{Y}{\sqrt{x_c^2 + v^2}}\right) + \tan^{-1}\left(\frac{x}{r}\right)\right).$$

[6.3.2] Y is uniform on (a, b) and V is uniform on (u, w). Y and V are independent. Thus,

$$H_Z(z) = \Pr\{Y \leq zV\}$$
$$= \frac{1}{w - u}\int_u^w \Pr\{Y \leq zv\}dv.$$

Furthermore,

$$\Pr\{Y \leq zv\} = \begin{cases} 0, & \text{if } v \leq \frac{a}{z} \\ \frac{zv - a}{b - a}, & \text{if } \frac{a}{z} < v \leq \frac{b}{z} \\ 1, & \text{if } v > \frac{b}{z}. \end{cases}$$

We distinguish the following cases:

(i) $\frac{b}{z} < u$, then

$$H_z(z) = \frac{1}{w - u}\int_u^w 1dv = 1$$

(ii) $\frac{a}{z} < u \leq \frac{b}{z} < w$, then

$$H_Z(z) = \frac{1}{w - u}\int_u^{\frac{b}{z}}\left(\frac{zv - a}{b - a}\right)dv + \frac{1}{w - u}\int_{bz}^w 1dv$$
$$= \frac{z}{2(w - u)(b - a)}\left(\left(\frac{b}{z}\right)^2 - u^2\right) - \frac{a}{(w - u)(b - a)}\left(\frac{b}{z} - u\right)$$
$$+\ (w - b/z)/(w - u).$$

(iii) $\frac{a}{z} \le u < w \le \frac{b}{z}$, then

$$H_Z(z) = \frac{1}{w-u}\int_u^w \left(\frac{zv-a}{b-a}\right) dv$$
$$= \frac{z}{2(w-u)(b-a)}(w^2-u^2) - \frac{a}{b-a}$$
$$= \frac{1}{b-a}\left(\frac{z(w+u)}{2} - a\right).$$

(iv) $u \le \frac{a}{z} < \frac{b}{z} \le w$,

$$H_Z(z) = \frac{1}{w-u}\int_{a/z}^{b/z} \frac{zv-a}{b-a} dv$$
$$= \frac{1}{(w-u)(b-a)}\left[\frac{z}{2}\left(\left(\frac{b}{z}\right)^2 - \left(\frac{a}{z}\right)^2\right) - a\left(\frac{b}{z} - \frac{a}{z}\right)\right].$$

(v) $u \le \frac{a}{z} < w \le \frac{b}{z}$,

$$H_Z(z) = \frac{1}{w-u}\int_{a/z}^{w} \frac{zv-a}{b-a} dv$$
$$= \frac{1}{(b-a)(w-u)}\left[\frac{1}{2}\left[w^2 - \left(\frac{a}{z}\right)^2\right] - a\left(w - \left(\frac{a}{z}\right)\right)\right].$$

Finally,
(vi) if $\frac{a}{z} > w$, then $H_Z(z) = 0$. All these results can be summarized as in Eq (6.8).

[6.3.3] Set $S(z,x) = \eta$, for $\eta > x$, then

$$\frac{\eta}{r} = \tan\left(2\sin^{-1}\left(\frac{z}{(1+(\frac{x}{r}+z(1+\frac{x}{r})^2)^{1/2})^2)^{1/2}}\right) + \tan^{-1}\left(\frac{x}{r}\right)\right).$$

Equivalently,

$$\frac{1}{2}\left(\tan^{-1}\left(\frac{\eta}{r}\right) - \tan^{-1}\left(\frac{x}{r}\right)\right) = \sin^{-1}(\cdot),$$

or

$$\psi(\eta) = \frac{z}{(1+(\frac{x}{r}+z(1+(\frac{x}{r})^2)^{1/2})^2)^{1/2}}.$$

Solving this equation for z, we obtain Eq. (6.9).

[6.3.4] The c.d.f. of Z for the given parameters is given in the following table

z	$H_Z(z)$
0.022	0.045455
0.023	0.092500
0.024	0.140000
0.025	0.187500
0.026	0.235000
0.027	0.282500
0.028	0.330000
0.029	0.377500
0.030	0.425000
0.031	0.472500
0.032	0.520000
0.033	0.567500
0.034	0.615000
0.035	0.662500
0.036	0.710000
0.037	0.757500
0.038	0.805000
0.039	0.852500
0.040	0.900000
0.041	0.941402
0.042	0.971190
0.043	0.990174
0.044	0.999091

From this table we obtain that the 10th, 50th and 90th percentiles of Z are

$$z_{.10} = .0235, \quad z_{.50} = .0315 \quad \text{and} \quad z_{.90} = 0.040.$$

Substituting these values in Eq. (6.7) we obtain $(U_s(x))_{.10} = 12.063$, $(U_s(x))_{.50} = 14.475$ and $(U_s(x))_{.90} = 17.042$.

[6.3.5] Using program SNGSHDW we obtain, for the field parameters of the previous exercise the following values of $Q_s(\eta \mid x)$.

η	$Q_s(\eta \mid 5)$	η	$Q_s(\eta \mid 5)$
11.00973	0.000000	15.15941	0.604043
11.47081	0.027249	15.62049	0.676578
11.93189	0.095514	16.08156	0.749083
12.39296	0.168238	16.54264	0.821556
12.85404	0.240937	17.00371	0.893996
13.31511	0.313612	17.46479	0.954601
13.77619	0.386261	17.92586	0.989059
14.23726	0.458883	18.38694	1.000000
14.69834	0.531477		

Here $U_m(5) = 11.00973$ and $U_M(5) = 18.38694$. We approximate $E\{U_s(x)\}$ by

$$U_m(5) + \Delta \sum_{j=1}^{16} [1 - Q_s(U_m(5) + j\Delta \mid 5)] = 14.5088,$$

where $\Delta = 0.46108$.

[6.3.6] The length of a shadow arc on $\mathcal{C}$, cast by a disk centered at (ρ, θ) with radius Y is $L(\rho, Y) = 2r \sin^{-1}(\frac{Y}{2\rho})$ (see Eq. (2.39)). Notice that the shadow length does not depend on the orientation angle θ. Assuming a standard Poisson field, and a uniform distribution of Y on (a, b), the p.d.f. of ρ on (U, W) is $f(\rho) = \frac{2\rho}{w^2 - u^2}$. Thus,

$$\begin{aligned} \Pr\{L(\rho, Y) \leq l\} &= \Pr\left\{Y \leq 2\rho \sin\left(\frac{l}{2r}\right)\right\} \\ &= \frac{2}{w^2 - u^2} \int_u^w \rho \Pr\left\{Y \leq 2\rho \sin\left(\frac{l}{2r}\right)\right\} d\rho. \end{aligned}$$

Furthermore,

$$\Pr\left\{Y \leq 2\rho \sin\left(\frac{l}{2r}\right)\right\} = \begin{cases} 0, & \text{if } \rho \leq \dfrac{a}{2\sin(\frac{l}{2r})} \\[2ex] \dfrac{2\rho \sin(\frac{l}{2}) - a}{b - a}, & \text{if } \dfrac{a}{2\sin(\frac{l}{2r})} < \rho \leq \dfrac{b}{2\sin(\frac{l}{2r})} \\[2ex] 1, & \text{if } \rho > \dfrac{b}{2\sin(\frac{l}{2r})} \end{cases}$$

Substituting this in the above integral we obtain the required result.

[6.5.1] For the given parameters we obtain $D_T(0 \mid -15) = .5266$. Hence $p_0 = 0.4734$.

[6.5.2] For the parameters of the problem we obtain the following values of the c.d.f. $D_T(t \mid x)$

t	$D_T(t \mid 0)$	t	$D_T(t \mid 0)$
7.68	0.00000	19.68	0.88425
8.68	0.29906	20.68	0.88425
9.68	0.29947	21.68	0.89479
10.68	0.30039	22.68	0.90462
11.68	0.30039	23.68	0.94825
12.68	0.30191	24.68	0.95689
13.68	0.69297	25.68	0.95689
14.68	0.69969	26.68	0.96454
15.68	0.70755	27.68	0.97109
16.68	0.70755	28.68	0.97663
17.68	0.71599	29.68	0.97663
18.68	0.87342	30.68	0.98127

From these results we obtain that the mean and standard deviation of L are, respectively, 14.379 and 5.269.

[6.6.1] (i) The probability for at most one shadow is $p_0 + p_1 = 0.038$.
(ii) $1 - (p_0 + p_1 + p_2) = 0.5934$.

[6.7.1] Program SURVFUC yields the results

j	$\hat{S}(20-j, 20)$
0	1.0000
1	0.8051
2	0.6444
3	0.5359
4	0.4512
5	0.3735
6	0.3244
7	0.2841
8	0.2402
9	0.2148
10	0.1924
11	0.1657
12	0.1488
13	0.1334
14	0.1188
15	0.1052
16	0.0936
17	0.0829
18	0.0732
19	0.0650
20	0.0575

References

1. Ailam, G. (1966), Moments of coverage and coverage spaces, **J. Appl. Prob. 3**, 550-555.
2. Ailam, G. (1970), On probability properties of measures of random sets and the asymptotic behavior of empirical distribution functions, **J. Appl. Prob. 5**, 196-202.
3. Aldous, D. (1989), **Probability Approximations via the Poisson Clumping Heuristic**, Springer Verlag, Applied Mathematical Sciences, **77**, New York.
4. Ambartzumian, R.V. (1982), **Combinatorial Integral Geometry, with Applications to Mathematical Stereology**, John Wiley, New York.
5. Ambartzumian, R,V, (1990), **Factorization Calculus and Geometric Probability**, Cambridge University Press, Cambridge.
6. Baddeley, A. (1982), Stochastic Geometry: An Introduction and Reading -List, **International Statistical Review**, **50**, 179-193.
7. Brémaud, P. (1981), **Point Processes and Queues**, Springer Verlag, New York.
8. Chernoff, H. and Daly, J.F. (1957), The distribution of shadows, **J. Math. Mech. 6**, 567-584.
9. Cooke, P.J. (1974), Bounds for coverage probabilities with applications to sequential coverage problems, **J. Appl. Prob.**, **11**, 281-293.
10. Cox, D.R. and Isham, V. (1980), **Point Processes**, Chapman and Hall, London.
11. Davy, P.J. (1982), Coverage, **Encyclopedia of Statistical Sciences**, Eds. S. Kotz and N.L. Johnson, John Wiley, New York.
12. Domb, C. (1947), The problem of random intervals on a line , *Proc. Cambridge Phil. Soc.* **43**, 329-341.
13. Dvoretsky, A. (1956), On covering a circle by randomly placed arcs, **Proc. Nat. Acad. Sci.** USA, **42**, 199-203.
14. Eckler, A.R. (1969), A survey of coverage problems associated with point and area targets, **Technometrics**, **11**, 561-589.
15. Eckler, A.R. and Burr, S.A. (1972), **Mathematical Models of target Coverage and Missile Allocation**, Military Operations Research Society, Arlington, Virginia.
16. Fava, N.A. and Santaló, L.A. (1978), Plate and line segment processes, **J. Appl. Prob.**, **15**, 494-501.
17. Feller, W. (1968), **An Introduction to Probability Theory and Its Applications**, Vol. I. Third Edition. Wiley, New York.
18. Flatto, L. (1973), A limit theorem for random covering of a circle, **Israel J. Math.**, **15**, 167-184.
19. Flatto, L. and Newman, D.J. (1977), Random covering, **Acta Math.**, **138**, 241-264.
20. Gilbert, E.N. (1965), The probability of covering a sphere with N circular caps, **Biometrika**, **52**, 323-330.
21. Glaz, J. and Naus, J. (1979), Multiple coverage of the line, **Ann. Prob.**, **7**, 900-906.
22. Greenberg, I. (1980), The moments of coverage of a linear set, **J. Appl. Prob. 17**, 865-868.
23. Guenther, W.C. and Terrango, P.J. (1964), A review of the literature on a class of coverage problems, **Ann. Math. Statist.**, **35**, 232-260.

24. Hafner, R. (1972), The asymptotic distribution of random clumps, **Computing, 10,** 335-351.
25. Hall, P. (1986) Clump counts in a mosaic, **Ann. Prob.**, **14**, 424-458.
26. Hall, P. (1988), **Introduction to the Theory of Coverage Processes**, Wiley, New York.
27. Holst, L. (1980), On multiple covering of a circle with random arcs. **J. Appl. Prob.**, **17**, 284-290.
28. Holst, L. and Hüsler, J. (1984), On the random coverage of the circle, **J. Appl. Prob.**, **21**, 558-566.
29. Hüsler, J. (1982), Random coverage of the circle and asymptotic distributions, **J. Appl. Prob**. **19**, 578-587.
30. Jarnagin, M.P. (1966), Expected coverage of a circular target by bombs all aimed at the center, **Oper. Res.**, **14**, 1139-1143.
31. Karlin, S. and Taylor, H.M. (1975), **A First Course in Stochastic Processes,** 2nd Edition, Academic Press, New York.
32. Karr, A.F. (1986), **Point Processes and Their Statistical Inference**, Marcel Dekker, New York.
33. Kellerer, A.M. (1983), On the number of clumps resulting from the overlap of randomly placed figures in a plane, **J. Appl. Prob**., **20**, 126-135.
34. Kendall, M.G. and Moran, P.A.P. (1963), **Geometrical Probability**, Griffin, London.
35. Matheron, G. (1975), **Random Sets and Integral Geometry**, John Wiley, New York.
36. Miles, R.E. (1964a), Random polygons determined by random lines in a plane, I. **Proc. Natl. Acad. Sci**. USA, **52**, 901-907.
37. Miles, R.E. (1964b), Random polygons determined by random lines in a plane, II. **Proc. Natl. Acad. Sci**. USA, **52**, 1157-1160.
38. Miles, R.E. (1971), Poisson flats in Euclidean spaces. Part II: Homogeneous Poisson flats and the complementary theorem, **Adv. Appl. Prob**. **3**, 1-43.
39. Miles, R.E. (1972), The random division of space. **Suppl. Adv. Appl. Prob.**, **4**, 243-266.
40. Miles, R.E. (1980), A survey of geometrical probability in the plane, with emphasis on stochastic image modeling. **Comp. Vision, Graph., Image Process**., **12**, 1-24.
41. Moran, P.A.P. (1973), The random volume of interpenetrating spheres in space. **J. Appl. Prob.**, **10**, 483-490.
42. Naus, J.I. (1979), An indexed bibliography of clusters, clumps and coincidences, **Int. Statist. Rev**., **47**, 47-78.
43. Ramalhoto, M.F. (1984), Bounds for the variance of the busy of the busy period of the $M/G/\infty$ queue, **Adv. Appl. Prob.**, **16**, 929-932.
44. Ripley, B.D. (1976), The foundations of stochastic geometry, **Ann. Prob.**, **4**, 995-998.
45. Roach, S.A. (1968), **The Theory of Random Clumping**, Griffin, London.
46. Robbins, H. E. (1944), On the measure of random sets, **Ann. Math. Statist**. **15**, 70-74.
47. Robbins, H. E. (1945), On the measure of random sets, II, **Ann. Math. Statist**. **16**, 342-347.

48. Rogers, C.A. (1964), **Packing and Covering**, Cambridge University Press, London.
49. Ross, S. (1976), **A First Course in Probability**, McMillan, New York.
50. Santaló, L.A. (1976), **Integral Geometry and Geometrical Probability**, Addison-Wesley, Reading, Massachusetts.
51. Schroeter, G. (1982), The variance of the coverage of a randomly located target by a salvo of weapons, **Naval Res.Log. Quart.**, **29**, 97-111.
52. Schroeter, G. (1984), Distribution of number of point targets killed and higher moments of coverage of area targets, **Naval Res. Log. Quart.**, **31**, 373-385.
53. Shepp, L.A. (1972), Covering the circle with random arcs, **Israel J. Math. 11**, 328-345.
54. Siegel, A. F.(1978a), Random space filling and moments of coverage in geometrical probability, **J. Appl. Prob. 15**, 340-355.
55. Siegel, A. F.(1978b), Random arcs on the circle, **J. Appl. Prob.**, **15**, 774-789.
56. Siegel, A.F. (1979), Asymptotic coverage distributions on the circle, **Ann. Prob. 7**, 651-661.
57. Siegel, A.F. and Holst, T. (1982), Covering the circle with random arcs of random size, **J. Appl. Prob. 19**, 373-381.
58. Solomon, H. (1978), **Geometric Probability**, SIAM, Philadelphia, PA.
59. Solomon, H. and Weiner, H. (1986), A review of the packing problem, **Commun. Statist.-Theor. Meth. 15**, 2571-2607.
60. Stadje, W. (1985), The busy period of the queueing systems $M/G/\infty$, **J. Appl. Prob.**, **22**, 697-704.
61. Stevens, W.L. (1939), Solution to a geometrical problem in probability, **Ann. Eugen.**, **9**, 315-320.
62. Stoyan, D. (1979), Applied Stochastic Geometry: A Survey, **Biometric J.**, **21**, 693-715.
63. Stoyan, D., Kendall, W.S. and Mecke, J. (1987), **Stochastic Geometry and Its Applications**, Akademie-Verlag, Berlin.
64. Takács, L.(1958), On the probability distribution of the measure of the union of random sets placed in a Euclidean space, **Ann. Univ. Sci. Budapest, Eotvos Sect. Math.**, **1**, 89-95.
65. Yadin, M. and Zacks, S. (1982), Random coverage of a circle with applications to a shadowing problem, **J. Appl. Prob.**, **19**, 562-577.
66. Yadin, M. and Zacks, S. (1985), The visibility of stationary and moving targets in the plane, subject to a Poisson field of shadowing elements. **J. Appl. Prob.**, **22**, 776-786.
67. Yadin, M. and Zacks, S. (1988), The distribution of measures of visibility on line segments in three dimensional spaces under Poisson shadowing processes, **Nav. Res. Logistics Quart.**, **35**, 555-569.
68. Yadin, M. and Zacks. S, (1986), Discretization of a semi-Markov shadowing process. Technical Report No. 2, Contract DAAG29-84-K-0191, U.S. Army Research Office.
69. Yadin, M. and Zacks, S. (1990). Multiobserver multitarget visibility probabilities for Poisson shadowing processes in the plane, **Nav. Res. Logistics Quart.**, **37**, 603-615.

70. Zacks, S. and Yadin, M. (1984), The distribution of the random lighted portion of a curve in a plane shadowed by a Poisson random field of obstacles, **Statistical Signal Processing**, Eds. E.J. Wegman and J.G. Smith, 273-286, Marcel Dekker, New York.

Appendix

Computer Programs

In the attached floppy disk there are executable programs, which are mentioned in the book. The programs can be found under subdirectories which correspond to the six chapters. The various programs were written either in QuickBasic or in Fortran. It is desirable to run the program in executable form (e.g. __________.EXE). Each program displays first what it computes, and the user is asked to insert the required parameters. The programs can be listed, read or printed by using an editor. All programs which are __________.BAS or __________.FOR are in ASCII form.

In the present appendix we provide a short description to the programs.

Chapter 1

Name of Program: DISTRIB

Description: This program calculates and displays the probability distribution/density function (p.d.f.) and the cumulative distribution function (c.d.f.) of several common probability distributions. First, select a distribution. Next, enter the parameters one at a time. Then choose the display options. NOTE: All responses to prompts must be in lower case characters.

Distribution Codes
binom – binomial distribution
hyper – hypergeometric distribution
pois – poisson distribution
norm – normal distribution
expon – exponential distribution
quit – exit the program

Chapter 2

No program is furnished.

Chapter 3

1. Name of Program: SIMVP

Description: This program computes the simultaneous visibility probability of n points. You will be asked to insert the field parameters. The angles s should be in degrees and inserted from the largest to the smallest.

2. Name of Program: VPANN

Description: The present program computes the simultaneous visibility probabilities for n points in the annular region. You will be asked to insert the parameters of the field. The angles should be inserted in degrees from largest to smallest.

3. Name of Program: VPTRA

Description: The present program computes the simultaneous visibility probabilities of n points in trapezoidal regions. You will be asked to insert the parameters of the field.

4. Name of Program: VPANNW

Description: The present program computes simultaneous visibility probabilities of several points, with windows. The field structure is annular. You will be asked to insert the field parameters. Angles, s should be in degrees, and inserted from largest to smallest.

Chapter 4

1. Name of Program: MOMTVPB

Description: The present program computes the simultaneous visibility probabilities of several targets and several observation points. You will be asked to insert the parameters of the field. The parameters of lines of sight, their order within the strips, the number of strips, and other parameters of interest can be found in output file "B:MOMTVPB.DAT".

2. Name of Program: MOMTVPW

Description: The present program computes the simultaneous visibility probabilities of windows around the target points. The setup is exactly like in program MOMTVPB.

3. Name of Program: THRDVPW

Description: The present program computes the simultaneous visibility probability of several points on a line segment in space. You will be asked to insert the field parameters.

Chapter 5

1. Name of Program: MOANNTSM

Description: The present program computes the first six moments of W. The computations are done by the simulation algorithm. The first line of the output are the moment of W, the second line are the mixed-beta approximation. The field structure is annular, standard Poisson.

2. Name of Program: MOANNREC

Description: The present program computes the first six moments of W. The computation is done by the recursive equations numerically. The first line of the output are the moment of W, the second line are the mixed-beta approximation. The field structure is annular, standard Poisson.

3. Name of Program: VPWALL

Description: The program computes simultaneous visibility probabilities of subsets of lines of sight, with windows around the target points. The field is standard Poisson. You will be asked to insert the values of the field parameters. The output is in file "B:vpwall.dat".

4. Name of Program: MOTRAP

Description: This program computes moments of visibility on a linear target curve in the two-dimensions, with windows. The field is standard Poisson. Disks are centered in a trapezoidal region. The target curve is a line segment of distance R from the origin. The region is bounded by lines of distance U and W from the origin.

5. Name of Program: VPNORM

Description: This program computes the simultaneous visibility probability for any number of targets, N. The obstacles field is specified between two straight lines U, V and consist of random disks with intensity AL. The location of the centers of disks follows a bivariate normal distribution. The diameter of a disk follows a uniform distribution on the interval (a, b).

6. Name of Program: VPNRMALL

Description: The program computes the simultaneous visibility probability for any subset of given five points. The obstacles field is specified between two straight lines U, W and consist of random disks with intensity AL. The location of the centers of disks follows a bivariate normal distribution. The diameter of a disk follows a uniform distribution on the interval (a, b).

7. Name of Program: VPANNIN

Description: The program computes the first two moments of W, for annular regions, according to the formulae of Problem [5.3.1]

Chapter 6

1. Name of Program: VIEWLNG

Description: The present program computes the conditional probabilities that a visible segment on a line, to the right of x has length at least 1. The output is printed also in file "B:viewlng.dat".

2. Name of Program: CDFSHDW

Description: This program computes the CDF of the right-hand limit of a shadow starting at a point x, which is cast on a straight line in a trapezoidal field. The output is printed also in file "B:shdwlgn.dat".

3. Name of Program: SURVFUC

Description: The present program computes the survival probabilities of a target moving along a straight line in a trapezoidal region. The output of this program is printed in file "B:survprb.dat".

4. Name of Program: DISTNSH

Description: The present program computes the CDF $W(y \mid x)$ of the right limit of a cycle length, and the probabilities of having 0, 1 or 2 shadows. The field structure is trapezoidal. The output is printed also in file "B:distnsh.dat".

5. Name of Program: SNGSHDW

Description: This program computes the CDF of the right-hand limit of a shadow starting at a point x, which is cast on a straight line in a trapezoidal field by a single disk. The output is printed also in file "B:shdwlgn.dat".

Index

Lecture Notes in Statistics

For information about Volumes 1 to 8
please contact Springer-Verlag

Vol. 9: B. Jørgensen, Statistical Properties of the Generalized Inverse Gaussian Distribution. vi, 188 pages, 1981.

Vol. 10: A.A. McIntosh, Fitting Linear Models: An Application of Conjugate Gradient Algorithms. vi, 200 pages, 1982.

Vol. 11: D.F. Nicholls and B.G. Quinn, Random Coefficient Autoregressive Models: An Introduction. v, 154 pages, 1982.

Vol. 12: M. Jacobsen, Statistical Analysis of Counting Processes. vii, 226 pages, 1982.

Vol. 13: J. Pfanzagl (with the assistance of W. Wefelmeyer), Contributions to a General Asymptotic Statistical Theory. vii, 315 pages, 1982.

Vol. 14: GLIM 82: Proceedings of the International Conference on Generalised Linear Models. Edited by R. Gilchrist. v, 188 pages, 1982.

Vol. 15: K.R.W. Brewer and M. Hanif, Sampling with Unequal Probabilities. ix, 164 pages, 1983.

Vol. 16: Specifying Statistical Models: From Parametric to Non-Parametric, Using Bayesian or Non-Bayesian Approaches. Edited by J.P. Florens, M. Mouchart, J.P. Raoult, L. Simar, and A.F.M. Smith, xi, 204 pages, 1983.

Vol. 17: I.V. Basawa and D.J. Scott, Asymptotic Optimal Inference for Non-Ergodic Models. ix, 170 pages, 1983.

Vol. 18: W. Britton, Conjugate Duality and the Exponential Fourier Spectrum. v, 226 pages, 1983.

Vol. 19: L. Fernholz, von Mises Calculus For Statistical Functionals. viii, 124 pages, 1983.

Vol. 20: Mathematical Learning Models — Theory and Algorithms: Proceedings of a Conference. Edited by U. Herkenrath, D. Kalin, W. Vogel. xiv, 226 pages, 1983.

Vol. 21: H. Tong, Threshold Models in Non-linear Time Series Analysis. x, 323 pages, 1983.

Vol. 22: S. Johansen, Functional Relations, Random Coefficients and Nonlinear Regression with Application to Kinetic Data, viii, 126 pages, 1984.

Vol. 23: D.G. Saphire, Estimation of Victimization Prevalence Using Data from the National Crime Survey. v, 165 pages, 1984.

Vol. 24: T.S. Rao, M.M. Gabr, An Introduction to Bispectral Analysis and Bilinear Time Series Models. viii, 280 pages, 1984.

Vol. 25: Time Series Analysis of Irregularly Observed Data. Proceedings, 1983. Edited by E. Parzen. vii, 363 pages, 1984.

Vol. 26: Robust and Nonlinear Time Series Analysis. Proceedings, 1983. Edited by J. Franke, W. Härdle and D. Martin. ix, 286 pages, 1984.

Vol. 27: A. Janssen, H. Milbrodt, H. Strasser, Infinitely Divisible Statistical Experiments. vi, 163 pages, 1985.

Vol. 28: S. Amari, Differential-Geometrical Methods in Statistics. v, 290 pages, 1985.

Vol. 29: Statistics in Ornithology. Edited by B.J.T. Morgan and P.M. North. xxv, 418 pages, 1985.

Vol 30: J. Grandell, Stochastic Models of Air Pollutant Concentration. v, 110 pages, 1985.

Vol. 31: J. Pfanzagl, Asymptotic Expansions for General Statistical Models. vii, 505 pages, 1985.

Vol. 32: Generalized Linear Models. Proceedings, 1985. Edited by R. Gilchrist, B. Francis and J. Whittaker. vi, 178 pages, 1985.

Vol. 33: M. Csörgo, S. Csörgo, L. Horváth, An Asymptotic Theory for Empirical Reliability and Concentration Processes. v, 171 pages, 1986.

Vol. 34: D.E. Critchlow, Metric Methods for Analyzing Partially Ranked Data. x, 216 pages, 1985.

Vol. 35: Linear Statistical Inference. Proceedings, 1984. Edited by T. Calinski and W. Klonecki. vi, 318 pages, 1985.

Vol. 36: B. Matérn, Spatial Variation. Second Edition. 151 pages, 1986.

Vol. 37: Advances in Order Restricted Statistical Inference. Proceedings, 1985. Edited by R. Dykstra, T. Robertson and F.T. Wright. viii, 295 pages, 1986.

Vol. 38: Survey Research Designs: Towards a Better Understanding of Their Costs and Benefits. Edited by R.W. Pearson and R.F. Boruch. v, 129 pages, 1986.

Vol. 39: J.D. Malley, Optimal Unbiased Estimation of Variance Components. ix, 146 pages, 1986.

Vol. 40: H.R. Lerche, Boundary Crossing of Brownian Motion. v, 142 pages, 1986.

Vol. 41: F. Baccelli, P. Brémaud, Palm Probabilities and Stationary Queues. vii, 106 pages, 1987.

Vol. 42: S. Kullback, J.C. Keegel, J.H. Kullback, Topics in Statistical Information Theory. ix, 158 pages, 1987.

Vol. 43: B.C. Arnold, Majorization and the Lorenz Order: A Brief Introduction. vi, 122 pages, 1987.

Vol. 44: D.L. McLeish, Christopher G. Small, The Theory and Applications of Statistical Inference Functions. vi, 124 pages, 1987.

Vol. 45: J.K. Ghosh (Editor), Statistical Information and Likelihood. 384 pages, 1988.

Vol. 46: H.-G. Müller, Nonparametric Regression Analysis of Longitudinal Data. vi, 199 pages, 1988.

Vol. 47: A.J. Getson, F.C. Hsuan, {2}-Inverses and Their Statistical Application. viii, 110 pages, 1988.

Vol. 48: G.L. Bretthorst, Bayesian Spectrum Analysis and Parameter Estimation. xii, 209 pages, 1988.

Vol. 49: S.L. Lauritzen, Extremal Families and Systems of Sufficient Statistics. xv, 268 pages, 1988.

Vol. 50: O.E. Barndorff-Nielsen, Parametric Statistical Models and Likelihood. vii, 276 pages, 1988.

Vol. 51: J. Hüsler, R.-D. Reiss (Editors), Extreme Value Theory. Proceedings, 1987. x, 279 pages, 1989.

Vol. 52: P.K. Goel, T. Ramalingam, The Matching Methodology: Some Statistical Properties. viii, 152 pages, 1989.

Vol. 53: B.C. Arnold, N. Balakrishnan, Relations, Bounds and Approximations for Order Statistics. ix, 173 pages, 1989.

Vol. 54: K.R. Shah, B.K. Sinha, Theory of Optimal Designs. viii, 171 pages, 1989.

Vol. 55: L. McDonald, B. Manly, J. Lockwood, J. Logan (Editors), Estimation and Analysis of Insect Populations. Proceedings, 1988. xiv, 492 pages, 1989.

Vol. 56: J.K. Lindsey, The Analysis of Categorical Data Using GLIM. v, 168 pages, 1989.

Vol. 57: A. Decarli, B.J. Francis, R. Gilchrist, G.U.H. Seeber (Editors), Statistical Modelling. Proceedings, 1989. ix, 343 pages, 1989.

Vol. 58: O.E. Barndorff-Nielsen, P. Blæsild, P.S. Eriksen, Decomposition and Invariance of Measures, and Statistical Transformation Models. v, 147 pages, 1989.

Vol. 59: S. Gupta, R. Mukerjee, A Calculus for Factorial Arrangements. vi, 126 pages, 1989.

Vol. 60: L. Györfi, W. Härdle, P. Sarda, Ph. Vieu, Nonparametric Curve Estimation from Time Series. viii, 153 pages, 1989.

Vol. 61: J. Breckling, The Analysis of Directional Time Series: Applications to Wind Speed and Direction. viii, 238 pages, 1989.

Vol. 62: J.C. Akkerboom, Testing Problems with Linear or Angular Inequality Constraints. xii, 291 pages, 1990.

Vol. 63: J. Pfanzagl, Estimation in Semiparametric Models: Some Recent Developments. iv, 112 pages, 1990.

Vol. 64: S. Gabler, Minimax Solutions in Sampling from Finite Populations. v, 132 pages, 1990.

Vol. 65: A. Janssen, D.M. Mason, Non-Standard Rank Tests. vi, 252 pages, 1990.

Vol. 66: T. Wright, Exact Confidence Bounds when Sampling from Small Finite Universes. xvi, 431 pages, 1991.

Vol. 67: M.A. Tanner, Tools for Statistical Inference: Observed Data and Data Augmentation Methods. vi, 110 pages, 1991.

Vol. 68: M. Taniguchi, Higher Order Asymptotic Theory for Time Series Analysis. viii, 160 pages, 1991.

Vol. 69: N.J.D. Nagelkerke, Maximum Likelihood Estimation of Functional Relationships. v, 110 pages, 1992.

Vol. 70: K. Iida, Studies on the Optimal Search Plan. viii, 130 pages, 1992.

Vol. 71: E.M.R.A. Engel, A Road to Randomness in Physical Systems. ix, 155 pages, 1992.

Vol. 72: J.K. Lindsey, The Analysis of Stochastic Processes using GLIM. vi, 294 pages, 1992.

Vol. 73: B.C Arnold, E. Castillo, J.-M. Sarabia, Conditionally Specified Distributions. xiii, 151 pages, 1992.

Vol. 74: P. Barone, A. Frigessi, M. Piccioni (Editors), Stochastic Models, Statistical Methods, and Algorithms in Image Analysis. vi, 258 pages, 1992.

Vol. 75: P.K. Goel, N.S. Iyengar (Editors), Bayesian Analysis in Statistics and Econometrics. xi, 410 pages, 1992.

Vol. 76: L. Bondesson, Generalized Gamma Convolutions and Related Classes of Distributions and Densities. viii, 173 pages, 1992.

Vol. 77: E. Mammen, When Does Bootstrap Work? Asymptotic Results and Simulations. vi, 196 pages, 1992.

Vol. 78: L. Fahrmeir, B. Francis, R. Gilchrist, G. Tutz (Editors), Advances in GLIM and Statistical Modelling: Proceedings of the GLIM92 Conference and the 7th International Workshop on Statistical Modelling, Munich, 13-17 July 1992. ix, 225 pages 1992.

Vol. 79: N. Schmitz, Optimal Sequentially Planned Decision Procedures. xii, 209 pages, 1992.

Vol. 80: M. Fligner, J. Verducci (Editors), Probability Models an Statistical Analyses for Ranking Data. xxii, 306 pages, 1992.

Vol. 81: P. Spirtes, C. Glymour, R. Scheines, Causation, Predic tion, and Search. xxiii, 526 pages, 1993.

Vol. 82: A. Korostelev and A. Tsybakov, Minimax Theory c Image Reconstruction. xii, 268 pages, 1993.

Vol. 83: C. Gatsonis, J. Hodges, R. Kass, N. Singpurwalla (Editors Case Studies in Bayesian Statistics. xii, 437 pages, 1993.

Vol. 84: S. Yamada, Pivotal Measures in Statistical Experimen and Sufficiency. vii, 129 pages, 1994.

Vol. 85: P. Doukhan, Mixing: Properties and Examples. xi, 14 pages, 1994.

Vol. 86: W. Vach, Logistic Regression with Missing Values in t Covariates. xi, 139 pages, 1994.

Vol. 87: J. Møller, Lectures on Random Voronoi Tessellations. v 134 pages, 1994.

Vol. 88: J.E. Kolassa, Series Approximation Methods Statistics. viii, 150 pages, 1994.

Vol. 89: P. Cheeseman, R.W. Oldford (Editors), Selecting Mod From Data: Artificial Intelligence and Statistics IV. x, 487 pag 1994.

Vol. 90: A. Csenki, Dependability for Systems with a Partition State Space: Markov and Semi-Markov Theory and Computatio Implementation. x, 241 pages, 1994.

Vol. 91: J.D. Malley, Statistical Applications of Jordan Algeb viii, 101 pages, 1994.

Vol. 92: M. Eerola, Probabilistic Causality in Longitudi Studies. vii, 133 pages, 1994.

Vol. 93: Bernard Van Cutsem (Editor), Classification Dissimilarity Analysis. xiv, 238 pages, 1994.

Vol. 94: Jane F. Gentleman and G.A. Whitmore (Editors), C Studies in Data Analysis. viii, 262 pages, 1994.

Vol. 95: Shelemyahu Zacks, Stochastic Visibility in Rand Fields. x, 175 pages, 1994.